PLANET PALM

PLANET PALM

HOW PALM OIL ENDED UP
IN EVERYTHING—AND
ENDANGERED THE WORLD

JOCELYN C. ZUCKERMAN

THE
NEW
PRESS

NEW YORK
LONDON

Requests for permission to reproduce selections from this book should be made through our website: https://thenewpress.com/contact.

Portions of Chapters 5 and 7 appeared in different forms in *Audubon* and *The Nation*, respectively, in articles commissioned and produced in collaboration with the Food & Environment Reporting Network. Portions of Chapter 8 appeared in different form in *Men's Journal*. A complete list of text and photograph permissions can be found on page 315.

Published in the United States by The New Press, New York, 2021
Distributed by Two Rivers Distribution

ISBN 978-1-62097-523-7 (hc)
ISBN 978-1-62097-524-4 (ebook)
CIP data is available

The New Press publishes books that promote and enrich public discussion and understanding of the issues vital to our democracy and to a more equitable world. These books are made possible by the enthusiasm of our readers; the support of a committed group of donors, large and small; the collaboration of our many partners in the independent media and the not-for-profit sector; booksellers, who often hand-sell New Press books; librarians; and above all by our authors.

www.thenewpress.com

Book design and composition by Bookbright Media
This book was set in Janson Text

Printed in the United States of America

10 9 8 7 6 5 4 3 2 1

For the keepers of the forests

For a colonized people, the most essential value,
because it is the most meaningful, is first and foremost the land:
the land, which must provide bread and, naturally, dignity.

—Frantz Fanon, *The Wretched of the Earth*

CONTENTS

PLANET PALM

PROLOGUE
Oil Crisis

KURT COBAIN was shrieking in my ears. I was rattling over a dirt road in a remote corner of southeastern Liberia, staring out the window of a Land Cruiser at a landscape rendered entirely in burnt orange. Just a few weeks earlier, the place had been dense forest, raucous with the chirps and squawks of birds, the scratching of animals in the underbrush. Clear streams had trickled over rocks. For generations, the families in this pocket of the West African nation founded by freed American slaves had collected rattan from the forest for building their houses and furniture. The men had returned in the evenings bearing honey, crabs, and groundhogs. Women, their infants tethered to their backs, had bent over plots of yams, melons, and beans in clearings by their huts.

All that remained of such time-honored tableaus now were the thousands of dead trees laid at intervals along the endless expanse of dirt. In the early-morning fog, they evoked fallen soldiers on a still-smoking battlefield. We drove on for miles, past a view comprising only scarred dirt and dead vegetation, punctuated by the occasional bright-yellow CAT excavator. The destruction, in both its scope and its finality, was like nothing I'd ever seen. And so had come the pounding drums and sneering guitars of Nirvana's "Rape Me," the grim earworm that would become the soundtrack to my trip. The more I saw, the louder the internal rage.

I'd driven down to Sinoe County from the Liberian capital, Monrovia, accompanied by an Italian photographer and a couple of local researchers, to report a story about land grabs, or large-scale territorial acquisitions by outsiders. The phenomenon, which had come to the world's attention in the aftermath of the financial and food crises of 2008, entails investment banks, pension funds, land-poor countries, and agribusiness seizing vast swathes of fertile ground in places like Ethiopia and Madagascar—places where traditional land rights are easy to exploit. I'd chosen to focus on Liberia in part because I'd long admired its president, Ellen Johnson Sirleaf, but also because of its historical ties to the United States. It wasn't until I got down on the ground that I discovered the grabbing had almost entirely to do with palm oil. And despite having spent twelve years on the staff of *Gourmet* and written dozens of articles about the environment and agriculture, I knew next to nothing when it came to this substance.

Liberia proved a rude awakening. The violence on display there extended beyond the destruction of the landscape to the Liberians themselves. In one village, a scattering of mud-block and thatch houses located inside an oil-palm concession run by a Singapore-based company, a fifty-year-old father of seven described how the outsiders had shown up and bulldozed the town in which he'd spent his entire life. Others talked of how the company had destroyed their crops and gravesites, polluted their streams, and run them out of their homes. "What I've lost is plenty," a fifty-three-year-old woman told me through tears. "We can't plant plantains. We can't plant rice. We can't plant peppers." The people who had ripped out her crops to replace them with oil palm had given the woman a one-time payment of a few hundred dollars. "It's finished a long time ago," she told me. A Monrovia-based lawyer advocating for the locals lamented the community members' loss of identity. "The guy who was a respected farmer," he said, "has now become a slave laborer."

The company responsible for the changes had been in operation in Sinoe for just thirteen months. It had signed an 865,000-acre lease good for sixty-five years, with an option for a thirty-three-year extension. In other words, it was just getting started.

I ended up not going with the "Rape Me" thing for the start of my article—I could picture my editors rolling their eyes—but the sick-making feeling from that trip stayed with me long after I'd filed my piece. The whole experience had hit weirdly close to home. While I may have been clueless about palm oil, I do know something about life on the equator in a remote African village. In my twenties, I spent two years working as a Peace Corps volunteer in western Kenya. I'd taught English and math to high school students in a tiny outpost hundreds of miles from the nearest khaki-clad tourist. There was no electricity or running water in the hamlet known as Buhuyi, and back then there were no cellphones, either. I would wake with the roosters, heat a pan of water over a camp stove for a splash bath, breakfast on papaya and milky tea, and then hop on my bike for the ride down the orange-dirt road to school. In the evenings, I'd compose long letters by candlelight or settle in under the mosquito net with a flashlight and a novel. It was lonely at times, and I came down with malaria twice, but in many respects those twenty-seven months were the happiest of my life. I loved the sound of the rain on my corrugated-tin roof, and the smell of the mud drying in the equatorial sun. I loved the slow rhythm of the days, the lack of pretense in my exchanges with the locals, the absence of extraneous anything.

Recently cleared land for oil-palm development in Sinoe County, Liberia.

Of course, there was also deep poverty in that village, and plenty
of frustration over the dearth of opportunity and the snail-like pace
of change. (It's possible I loved it so much only because I knew I had
a ticket out.) But the people of Buhuyi had neat farms and close-knit
families. They had rich soil, mango and jackfruit trees, maybe a cow
or some chickens. The air was fresh, and the rivers ran clear. We all
laughed a lot. In the aftermath of my reporting trip, I was waking
from nightmares about having returned to Buhuyi to find that all my
students' farms had been replanted with oil palm. My village looked
like that horrible corner of Liberia.

But don't get me wrong: I know that nothing about international development is straightforward. I've also lain awake wonder-
ing whether I'd done that Liberia story justice. Had my visceral
anger been misplaced? Had mine been the naive take of your typical parachuting-in journalist (one with health insurance and a gym
membership back home in Brooklyn)? Certainly the Liberia that I'd

observed outside of those plantations was no Disney theme park. The road we'd taken from Monrovia, for instance, had been so riven with ditches, potholes, and impromptu lakes that it had taken us eight hours to go a mere 150 miles. Sinoe County, though home to more than one hundred thousand people, had the aura of a place forgotten to time. "This is 'the city'?" the photographer had remarked as we finally rolled into its capital. "This is a shantytown."

By the time I arrived in Liberia, the country was more than a decade out from the brutal civil wars that had engulfed it from 1989 to 2003, taking some 250,000 lives in the process. (The Ebola outbreak wouldn't happen for another year, when it would nearly kill one of those researchers in the Land Cruiser, but that's a story for another time.) Even so, the place was in bad shape. What little success Johnson Sirleaf had achieved with its economy had come in the form of concession agreements signed with outsiders drawn to the country's natural wealth (see: land grabs), which, in addition to farmland, includes rich timber and mineral resources. I know well that the World Bank ranks agricultural growth high on its list of priorities for countries looking to reduce poverty—a point echoed by all of the palm oil executives I'd spoken with for my story. Liberia had been blessed with fertile land but was in desperate need of infrastructure and jobs, they said, challenges that they were in a position to address. Sure, there might be some growing pains along the way, but in the end, oil-palm plantations could well prove the country's salvation.

But could they really? Had they done so anywhere else? And just what *was* palm oil, anyway, and how was it possible that the world suddenly needed so much of this stuff? In the months following that Liberia trip, I became fairly obsessed with figuring all of this out. And I discovered that the landscape overhaul taking place in Liberia was already well underway in Southeast Asia, and that the result of this agricultural revolution is literally everywhere. In the space of just a few decades, palm oil has quietly insinuated itself into every facet of our lives, with roughly half of all products in U.S. grocery stores now containing some part of the plant. (Though the commodity in question is palm oil, the plant from which it derives—it's not technically a tree—is called the *oil palm*.) Palm oil alone now counts for one-third of total global vegetable-oil consumption.

When you wake up in the morning and brush your teeth? Palm oil in the toothpaste. Step into the shower? Palm oil in the soap, the shampoo, and the conditioner. Ditto the moisturizer, mascara, and lipstick you might apply afterward. Down in the kitchen, there's palm oil in the non-dairy creamer, in the doughnuts, in the baby formula, and in the dog food. It's in the Nutella you spread on the kids' toast. (If you bought the bread at the supermarket, it's likely in that, too.) Palm oil in the crackers and ice cream at lunch, in the Snickers or Cheez-Its you might nibble mid-afternoon. At dinner, it's more of the same—including in the feed consumed by whatever piece of cow, pig, sheep, or chicken now sits on your plate.

If Americans seem to be OD'ing on the stuff, it's even more extreme overseas. India, now the world's number-one importer of palm oil, went from buying 30,000 metric tons in 1992 to *9.2 million* in 2019. China saw an increase from 800,000 metric tons to 6.4 million over the same period. Worldwide, production of palm oil has more than doubled in just the last fifteen years; oil-palm plantations now cover more than 104,000 square miles—an area larger than New Zealand. With producers running out of land in Indonesia and Malaysia, which together account for some 85 percent of today's global palm oil supply, they're expanding to Papua New Guinea, the Philippines, and the Solomon Islands, and farther afield to Latin America, as well as, obviously, to Africa. Last year, global consumption reached nearly

72 million metric tons—that's roughly twenty pounds of palm oil for every person on the planet.

"You're soaking in it," went the old tagline of the palm oil–based dish detergent Palmolive. Little did the admen behind the campaign, which debuted in 1966, know how prescient the words of Madge the manicurist would one day prove: in the five decades since, palm oil imports to the United States have increased from 29,000 metric tons to more than 1.5 million. In the last fifteen years alone, imports to this country have risen a whopping 263 percent, thanks in part to the Food and Drug Administration's ban on trans fats. Semi-solid at room temperature, palm oil emerged as the ideal swap-in for the partially hydrogenated oils formerly used to enhance the texture and extend the shelf life of products like cookies and crackers. In addition to its widespread presence in processed foods, cosmetics, and personal-care products, palm oil is used in all sorts of industrial materials and, increasingly, as a biofuel.

How did a crop most Americans have barely heard of—let alone ever seen—come to permeate our lives so completely? And what about this substance has made it so suddenly indispensable to industries across the globe? Did the downtrodden villagers and pissed-off laborers I'd met in Liberia reflect oil-palm communities worldwide, or had I stumbled upon a particularly unfortunate few? What were the long-term environmental implications of this unprecedented agricultural boom? How about the impact that all of these new calories were having on our bodies? Had consumers really demanded this product, or had it somehow been the other way around?

In setting out to answer these questions, I embarked on what would become a yearslong inquiry, one that took me across four continents and back in time more than two centuries. This obscure plant, I dis-

covered, has played an outsize role in shaping the world as we know
it, from spurring the colonization of Nigeria and greasing the gears
of the Second Industrial Revolution to transforming the societies of
Southeast Asia and beyond. In the same way that salt, cotton, and
sugar have reshaped our economies and landscapes and reshuffled
our geopolitics and health concerns, so, too, has palm oil done—and
continues to do today. Following the plant's journey over the decades
has served as a sort of master class in everything from colonialism
and commodity fetishism to globalization and the industrialization of our modern food system. (I also learned a lot about makeup.)
Today, palm oil stands center stage in what *The Lancet* has termed
the Global Syndemic: the combined twenty-first-century crises of
obesity, malnutrition, and climate change.

But I'm getting ahead of myself. To tell this story properly, we need
to start at the beginning. Which, oddly enough, takes us right back
to Liberia, or to its general environs, anyway. The oil-palm plant,
Elaeis guineensis, is native to West and Central Africa, flourishing
along a curvy band that runs down the western coast of the continent
from Guinea to the Democratic Republic of Congo, where it swells
eastward to cover much of the middle of that country. A perennial
plant, it resembles the more familiar coconut palm in that it features
draping, oversized fronds. But instead of coconuts tucked under its
leafy canopy, you'll find spiky brown bunches cradling hundreds of
plum-sized, bright-orange fruits. "They were like dried and withered
heads, the product of a savage massacre," writes Graham Greene in a
telling passage from his 1960 novel *A Burnt-Out Case*, part of which
takes place on a Congolese oil-palm plantation.

Archaeological findings suggest that the Egyptians were trading
palm oil as early as 3,000 B.C.; in the fifth century B.C., the Greek
historian Herodotus reported the likely use of palm wine in the

preparation of mummies. When West and Central Africans began clearing land for agriculture, they would leave a few oil palms standing, prizing them for use in everything from cooking and winemak-

Harvesting oil palm in West Africa, circa 1909.

ing to house construction and medicines. When the agriculturalists left those fields to lie fallow after one or two harvests, moving so that the soil and forest could regenerate, the oil palms would continue to grow, becoming a part of the emergent secondary forest. Fruits scattered by people and animals would propagate into still more oil palms, thereby extending the range of the species. It's for this reason that the oil-palm groves of the region are referred to as "sub-spontaneous."

The plant's shiny fruits actually bequeath two oils—one from the tangerine-colored pulp and another from the central kernel—each of which lends itself to numerous applications. (I've referred only to "palm oil" in the preceding paragraphs in order to keep things simple; from here on in, I'll distinguish between palm oil and palm-kernel oil.) The plants, which begin bearing fruit at about three years old and have an economic life of some twenty-five, are uncommonly productive, yielding considerably more oil per acre than either soy or rapeseed. In its unrefined form, palm oil is an excellent source of vitamins A and E.

In parts of rural Liberia—as in Cameroon, Nigeria, and across the oil-palm belt—achieving that unrefined form involves pretty much the same procedure it would have involved centuries ago. It begins with a guy (this part is still always done by a guy) scaling one of the skinny trunks to reach the ripe bunches at the top. Using a sling crafted from local vines, one section stretching around his body and the other lassoed to the tree, he leans back, feet planted on the trunk, and sort of shimmies his way up. With the trees reaching as high as ninety feet, falls can be fatal. Using a machete, he then hacks at the desired bunch until it goes hurtling to the ground. (Locals know to keep their distance, as a flying fifty-pound oil-palm bunch can do some serious damage.) It's at this point that the women get involved, "sweating" the fruits under a mat to loosen them from the spiky husks and then boiling them in metal drums to soften the fruit and slow the development of free fatty acids, which lead to rancidity. The steamed fruit then gets dumped into a vat to be stomped on, old-world-winemaker style, or to a mill, whether human-, animal-, or machine-powered, for crushing. The nuts having been removed (they'll be cracked later to attain the kernels inside), the resulting mustard-colored mash then gets transferred to a trough filled with water, where the oil floats to the surface to be skimmed off. After a final pass over the flames, the oil gets poured through a basket or other filter to remove any remaining fibers, leaving the brick-hued finished product.

Europeans sailing down the African coast would have encountered such home-processed oil, as well as kernels, for sale in the ports where they dropped anchor. "This palm oil is of great use to the inhabitants . . . in several respects," wrote a French merchant named John Barbot in 1732, "for besides its serving to season their meat, fish, etc., and to burn in their lamps to light them at night, it is an excellent ointment against rheumatick pains, winds and colds in the limbs, or other like diseases." It is, he added, "no despicable sauce, especially when new." (The African American culinary historian Jessica Harris notes that after standing for a few days, palm wine, made from the sap of the plant, "has the kick of the proverbial country mule and becomes western Africa's form of white lightning.")

In his 1958 classic *Things Fall Apart*, the Nigerian novelist Chinua Achebe writes that "proverbs are the palm oil with which words are eaten," one of the book's numerous references to the plant and its myriad derivatives. ("Those whose palm-kernels were cracked for them by a benevolent spirit," a village elder advises at one point, "should not forget to be humble.") By the 1890s, the era in which Achebe's novel takes place, a thriving international trade in palm oil was under way along the continent's western coast. Its heart, at the mouth of the Niger River, in today's southern Nigeria, became known as the Oil Rivers—a curious foreshadowing of the role that the Delta would play a century later as the troubled trading ground for a different sort of oil. Europeans originally sourced the oils for lighting their lamps, but the two substances would eventually find their way into soaps and candles, and into the lubricants required by the age's shiny new machinery. Tinplate makers came to prefer palm oil over tallow for their hot oil baths. Eventually, tins made with the oils would conserve Europeans' food, palm kernels would nourish their dairy cows, and, by the end of the nineteenth century, palm kernel oil would make its way into the faux butter they were spreading on their toast. But as Achebe's novel makes painfully clear, the societies long in place at the source of all that bounty would emerge from the trade far from unscathed. Whereas in 1870 more than 80 percent of Africa south of the Sahara was ruled by indigenous chiefs and kings, by 1910, all that had changed, with the region a patchwork of colonies, protectorates, and territories overseen by white newcomers.

The first part of this book traces this disturbing past, telling the early palm oil story through the lens of a handful of, well, quintessential men of empire. One of them, a to-the-manor-born maverick named George Goldie, would ultimately be credited with having secured the Nigerian colony for the Crown. William Lever, a quirky shopkeeper's son, would go on to found what would become the multinational conglomerate Unilever. He followed in the footsteps of King Leopold II to establish oil-palm plantations in the Belgian Congo, adopting not a few of the infamous monarch's murderous ways.

We also follow the fruit across the Atlantic on slave-bearing vessels destined for the sugar and tobacco plantations then being established by the Portuguese in Brazil. En route, palm oil was used for feeding the captives, and the kernels eventually found their way into the New World dirt, thus establishing there the sort of sub-spontaneous groves common to Africa. Escaped and freed slaves built communities along the coast amid those groves, adapting the culinary and spiritual traditions of their African forbears as a form of resistance and a way of preserving their identity. Today, *dendê*, as the oil is known in Brazil, features prominently not just in the region's traditional dishes, but in its religious ceremonies and in much of its art.

Finally, we head in the other direction, to Malaya and the Dutch East Indies, where we embed with a handful of "cultured" Brits, Frenchmen, Belgians, and Danes as they ship out, in all their enthusiastic youth, to seek adventure and fortune in the exotic east. The legacies—indeed, many of the very companies—left behind by such pioneers as Henri Fauconnier, Adrien Hallet, and Aage Westenholz continue to loom large over the palm oil industry today.

As to those massacres alluded to by Greene: you'll see that they're sadly dotted throughout this first section, for the palm oil business was nothing if not brutal, as the repeated and bloody confrontations will reveal.

"The Indians who labored in the plantations," the narrator of Tash Aw's 2019 novel *We, the Survivors* explains, "worked for the big corporations, the ones the government took over from the British. New owners, same rules. Times change but the workers' lives never improve. They had bad pay, bad housing, no schools, had to work with poisonous chemicals all day, had no entertainment in the evenings other than to drink their home-made *samsu* that made them go blind and mad."

In the second part of *Planet Palm*, I move on from the colonial era, doing my best to hand the narrative over to the sorts of invisible plantation workers and villagers that people Aw's portrait of his Malaysian homeland. First, we look at the impact of the industry on the indigenous tribes, smallholder farmers, and native animals of northern Sumatra, just across the Malacca Strait from Aw's benighted plantation laborers. Like most of the places targeted for oil-palm development—*Elaeis guineensis* thrives at 10 degrees to the north and south of the equator, a swathe that corresponds with the planet's tropical rainforests—Sumatra is a biodiversity hotspot. It is home to all manner of exotic birds, as well as to such iconic animals as the Sumatran elephant, rhinoceros, and tiger. A soft-spoken local takes me into the rainforest, where we brave mosquitoes and leeches to experience the world as it existed when ruled by those majestic creatures, and a guy who poaches critically endangered helmeted hornbills shows off his gun and his expert birdcalls. We also meet a

Modern-day artisanal oil producers in Liberia.

rowdy British primate specialist, who opens up about the challenges involved in trying to save the last of the world's Sumatran orangutans.

In a little concrete house in northwestern Honduras, I sit across from a thirty-four-year-old whose right arm stops just below the elbow and whose body is otherwise a patchwork of burns and skin grafts. While harvesting oil-palm fruits for the company Grupo Jaremar, Walter Banegas inadvertently tipped his aluminum harvesting pole onto an electric wire. (Those working on industrial plantations no longer have to scale the trees.) Jaremar's operations occupy land formerly owned by the American company United Fruit (later Chiquita), and the low wages, lack of healthcare, and general sense of insecurity that characterized the original Banana Republic continue to define the Central American industry today. Moving on from Honduras, I consider the state of labor across the palm oil economy, from smuggled migrant workers stripped of their passports (and their humanity) in Malaysia to child laborers in Indonesia and women exposed to sexual abuse and dangerous chemicals across three continents.

"People are using and selling a lot of palm oil here," a journalist in the Indian capital of New Delhi tells me, "but they don't talk about it." For Chapter 7, I travel to the world's number-one importer of the commodity to investigate what sort of impact the palm oil revolution is having on public health. Trade liberalization and economic growth in middle-income countries over the last two decades has led to a surge of oil flowing across international borders, where it's enabled the production of ever greater amounts of deep-fried snacks and ultra-processed foods. Rates of obesity, diabetes, and heart disease are soaring in the poorer countries where the multinational companies that peddle such junk are focused on growing their markets.

In 2015, an extended episode of haze linked to fires on oil-palm plantations in Indonesia led to an estimated one hundred thousand premature deaths. (A few weeks into the crisis, government officials ordered the evacuation of all babies under the age of six months.) As yet untallied is the long-term health damage caused by those conflagrations. The fires proved so difficult to extinguish in part because of the unique composition of the terrain on which so many of them burned. Indonesia is home to Earth's largest concentration of tropical peatlands—soils formed over thousands of years through the accumulation of organic matter—and when farmers and palm oil companies drain and burn that land as a precursor to planting, massive quantities of carbon dioxide escape into the atmosphere. Though many companies have signed zero-deforestation commitments and otherwise pledged to protect the environment, I learn while traveling undercover in Sumatra that fruit grown illegally on peatlands and other protected areas routinely makes its way to their mills and, ultimately, to our own kitchens, bathrooms, and fuel tanks. Much of the blame for this catastrophe can be laid at our own feet: In the mid-2000s, the United States and Europe put in place energy policies that failed to anticipate the devastating knock-on effects they would have on the other side of the globe.

At the end of 2018, after a U.K. grocery-store chain announced its intention to remove palm oil from its branded products in an effort to help save the rainforests, Malaysia's $16 billion industry responded with a social-media campaign personally attacking the company's managing director. When, a few months later, the European Union announced that it would phase out the use of palm oil in biofuels for similar reasons, Indonesia threatened to pull out of the Paris Climate Agreement. The last part of *Planet Palm* looks at these and other industry tantrums, and traces the political forces and dark money at work behind the scenes of the $65 billion business—from permits issued from inside jail cells and owners hidden behind offshore shell companies to long-dead villagers signing away their rights and elders hoodwinked by sweet-talking executives. In 2019, the World Health Organization compared the tactics used by the palm oil industry to those employed by the tobacco and alcohol lobbies, no slouches when it comes to playing dirty. Nor are these games restricted to foreign fields. It recently emerged that a Malaysian campaign accusing industry critics of being "neo-colonialists" was in fact the (highly compensated) work of a Washington, DC–based lobbying firm, one whose previous clients include, in addition to much of the Republican establishment, both Exxon and the former Burmese military junta.

Across the globe, those who've dared to speak out against the industry, whether laborers, peasant farmers, environmental activists, or investigative journalists, often have been met with violence. While writing this book, I fielded email and WhatsApp messages from folks in Sierra Leone, Honduras, Cameroon, Guatemala, the Democratic Republic of Congo, Peru, Colombia, Papua New Guinea, Indonesia, and Malaysia with updates on various protests, strikes, imprisonments, and murders, all of them related to palm oil. "There's an awful lot at stake," Nigel Sizer, the president of the Rainforest Alliance, which works with producers in Latin America and elsewhere, told me in his Manhattan office. "Massive investments in plantations. The infrastructure involved in processing. You get in the way of that, and you're going to be dealt with in the most brutal way imaginable."

It's a testament to human ingenuity, I suppose, that we've managed to transform a tiny fruit into endless configurations so as to satisfy our ever-changing desires. But in looking at the story of palm oil past and present, I think we need to consider at just what cost all of this progress has come. In 2019, hundreds of international experts issued a report finding that global biodiversity is declining faster than at any other time in human history, with one million species already facing extinction, many within decades, unless the world takes transformative action. Tropical rainforests, though they cover less than 10 percent of Earth's land surface, support more than half of the world's biodiversity. Some months ago, fires were again raging across Indonesia, giving off a toxic haze that turned the Sumatran sky an otherworldly shade of red. ("This is Earth," tweeted one freaked-out local, "not Planet Mars.") By the time they were extinguished, some 2,500 square miles of rainforest had been torched, and enough carbon had been spewed into the atmosphere to rival the annual output of Canada.

Navigating the way forward won't be easy. At one point in *We, the Survivors*, novelist Aw describes a busy Malaysian port town that's recently fallen on hard times: "You'd see the buses and the markets," he writes, "shopkeepers sweeping the pavements outside their doors, people sitting down at roadside food stalls—but you'd miss the feeling of anxiety, the knowledge that the entire town depended on trade from faraway places, goods being bought and sold by people we would never know. . . . The Europeans want to save the fucking planet so they ban the use of palm oil in food; within a month the entire port is on its knees."

Yet save the planet we must. And we don't have a remote chance of doing so if one of the key players in its demise is left to steamroll ahead, largely out of sight. As I wrap up this project, I've been returning in my mind to that original Liberian reporting trip, and I've decided that my anger wasn't in fact misplaced. The more I've learned about palm oil, the more convinced I've become that it's a substance that should concern us all. Because as I've ventured from

Kinshasa office park to Guatemalan river town, from Delhi street market to Bornean rainforest, bouncing over endless dirt roads and hailing all manner of sketchy-looking Ubers, with heroic fixers variously translating my questions into Bahasa Indonesia, Spanish, Kikongo, Portuguese, Liberian Kreyol, and Hindi, I've been repeatedly struck by the massive, massive reach of this industry—not just in stretching back centuries and impacting the lives of millions of people every day, but in the very real extent to which it will shape our common future.

PART I

♦

UNGUENT OF EMPIRE

1

Goldie Goes In

The elders consulted their Oracle and it told them that the strange man
would break their clan and spread destruction among them.

—Chinua Achebe, *Things Fall Apart*

IN MY mind's eye, George Dashwood Taubman Goldie is sipping a glass of absinthe at a sidewalk café on the Boulevard Haussmann when the news hits, on the evening of September 19, 1870, that Prussian soldiers have surrounded the French capital. It had been just a few weeks since the rangy twenty-four-year-old had convinced the family governess to run off with him to Paris, and now this? He doesn't yet realize just how bad his timing has been.

The fact is that Goldie, though raised in an aristocratic family, had been a screw-up for most of his life. After a coddled upbringing in a stone mansion overlooking the sea on the Isle of Man—his dad was descended from Edward I—he'd trained at the Royal Military Academy but lasted just two years in the army's Royal Engineers. "I was like a gun powder magazine," he would later tell a friend, adding that he'd been "blind drunk" when sitting for (and passing) his final exam.

Then there were the three years spent wandering the Egyptian Sudan with a young Arab woman he'd met after unexpectedly landing a family inheritance. He'd eventually left his "Garden of Allah" behind, only to lead "a life of idleness and dissipation" back in England. The Parisian adventure with the governess, a young woman named Mathilda Catherine Elliot, had been intended as a sort of fresh start. Instead, the young lovers would spend the next four months holed up and surviving, like the rest of the city during the historic siege, on the likes of dog and rat meat. They slunk back to England in February 1871 and were married a few months later. So when Goldie's eldest brother, John, confided to him in 1875 that he had just bailed out his father-in-law by purchasing the older man's near-bankrupt trading firm, Goldie jumped at the idea that he might be the one to save the thing. Holland, Jacques & Company had been sourcing palm oil from the West African coast since 1869.

Trade between Europe and West Africa was nothing new, of course. In the fifteenth century, the Portuguese had dropped anchor off the coast of what is now Ghana and begun exchanging cloth, iron, and copper for gold sourced from the continent's interior. Other European powers followed, and by the sixteenth century, the traffic in human beings was well under way. By 1792, some one hundred thousand enchained Africans were disembarking on the shores of the New World every year. Roughly half of them arrived on ships that had originally set sail from Liverpool.

Though Parliament outlawed the slave trade in 1807, the institution was foundational to Britain's commercial empire. Between 1750 and 1780, for example, some 70 percent of the government's total income came from taxes derived from its slave-powered colonies. (No one ever asks where all those young men looking for wives in Jane Austen novels got their money, a Ugandan friend noted recently.) The Niger Delta, in particular, spreading some two hundred seventy miles along the coast from Lagos to the Cameroonian border, had proven a lucrative hunting ground. Over the course of decades, Liverpool merchants had invested millions of pounds in ships and labor and built relationships with the African middlemen who coordinated the ferrying of people from the hinterland. (The internal traffic in humans had existed for centuries, though under a very different guise.) City-states such as Bonny, New Calabar, and Brass, situated along the low-lying plains of the Delta, had grown up around their slave markets.

Its wealth notwithstanding, the Delta was no French Riviera. "I

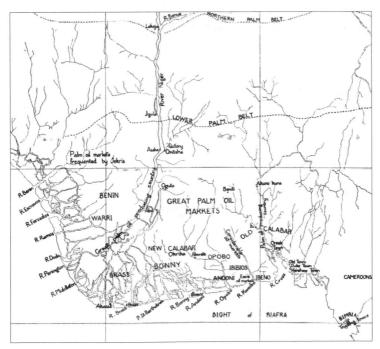

The Niger Delta circa 1870.

believe," wrote the explorer Mary Kingsley in her 1897 memoir *Travels in West Africa*, "the great swamp region of the Bight of Biafra [the eastern part of the Delta] is the greatest in the world, and that in its immensity and gloom it has a grandeur equal to that of the Himalayas."

She was being kind. The labyrinth of tidal creeks and inlets had a reputation as a ghastly place, all mangrove swamps and malarial menace. Europeans referred to the "unwholesome miasma" that emanated from the coast, where the mosquitos were fearsome and the soil too wet to grow much of anything. For the duration of the slave trade, the Europeans mostly kept to their ships offshore rather than brave the diseases and other unknowns lurking beyond the coastline. Upon his appointment as consul to the Bights of Benin and Biafra in 1861, the explorer Richard Burton had cursed the officials who'd determined his post. "They want me to die," he wrote to a friend, "but I intend to live, just to spite the devils."

Still, there were fortunes to be made, and for that the Europeans kept coming. They were none too pleased, then, when Britain began enforcing its slave ban with a naval blockade. The commerce in humans continued sporadically for years, finally dying out only after other countries enacted prohibitions of their own. In the meantime, the traders began turning their attention to the "legitimate" product that was palm oil. *Elaeis guineensis* grew wild in the forest belt behind the mangrove swamps.

With the Second Industrial Revolution transforming life back home in England, vegetable oil was suddenly in hot demand as a lubricant for the era's new railroads and machines. Palm oil also proved ideal as a tinning flux, and it was the basis for most of the candles being made at the time. An exploding population of workers, many of them traveling to gritty factory-line jobs in the shadow of belching smokestacks, fueled a new market for soap, which, thanks to the discoveries of a French chemist named Michel-Eugène Chevreul, had just begun to be mass-manufactured using vegetable oil. By 1850, Liverpool was producing some thirty thousand tons of palm oil–based soaps every year. At the same time, Manchester and other European towns were cranking out greater volumes of goods for exchange with Africa, from textiles woven with American-grown cotton to alcohol and cheap muskets.

By the time Goldie got in the game, the value of palm oil exports rivaled that of the slave trade at its peak, with the Niger Delta ground zero for the boom. "I have seen it written by a hand which sets down the truth in all things," wrote a lieutenant named John Hawley Glover in 1857, "that supercargoes [merchant-ship officers] then could make their 6,000 pounds a year in commission and yearly fortunes besides for the far-off owners of the time, and in a few years set up as merchants in their turn. A hundred pounds laid out in beads and coloured cloth brought thousands in the value of the oil, and no wonder that they called them the golden days of prosperity." Another John, one Sir Tobin of Liverpool, had funneled the profits he made from the early days of the palm oil trade toward an 1819 mayoral campaign in that city. He is said to have paid out six shillings for every vote in what was described by the local newspaper as one of "the most barefaced acts of bribery that ever disgraced even the electioneering annals of this venal rotten borough."

So potentially lucrative was the trade that it drew an entirely new class of merchants. Whereas most Europeans shipping out to the "Dark Continent" in those days tended to frame their missions in terms of explorer David Livingstone's "Three Cs," the men chasing palm oil made no pretenses toward Christianity or civilization. These guys were in it for the cash. Rude, uneducated types who drank and

George Goldie, detail from 1899 oil painting by H. Von Herkomer.

caroused with abandon, they resorted to violence to settle the tiniest
of disputes and well earned the nickname that would stick: "palm-oil
ruffians." Edward Nicolls, an officer serving in the Delta during the
1830s, complained of "the total want of principle amongst the greater

part" of this new gang. "There is no infamy or enormity that some of these Liverpool commanders will stop at."

Goldie was cut from a different cloth entirely, but his piercing blue eyes were similarly trained on the bottom line. And if Holland Jacques had failed to capitalize on the new commodity, he damn well didn't intend to. Say what they might about the guy's misspent youth, there was no question that he possessed a stellar mind. (Eight-year-old Goldie had been ejected from a "Calculating Man" performance on account of his having shouted all the answers to the audience-posed math problems before the purported genius on stage had figured them out.) Having finally stepped foot in the famous Delta mud, though, Goldie discovered that the situation was more complicated than he'd realized.

Looking on as scores of canoes shuttled back and forth to the ships anchored offshore, their African crews chanting as they paddled, he surmised that a relatively efficient system had been put in place. From 5:30 a.m. until well after three in the afternoon, crewmen heaved the oil-heavy barrels up on deck, where other workers, drenched in sweat despite the makeshift roofs of bamboo and palm fronds, pounded and tightened additional wooden casks or boiled and strained the viscous oil to preserve it for the journey back to Europe. Aside from the additional labor required up-country to convey the oil to the river's edge—unlike captives, the unwieldy barrels couldn't be force-marched across the land—the riparian highway network that had facilitated the traffic in people lent itself readily to the trade in oil.

The problem was with the folks in charge. The palm oil business, Goldie learned, operated under a system of barter known as "trust." The U.K. traders—most hailed from Liverpool, with a smattering from Bristol, London, and Glasgow—would offload their muskets, booze, and various useless trinkets to African middlemen and then hunker down offshore, sometimes for several months, as the latter ferried the goods inland and restocked their canoes with oil. Like the traders, the middlemen had pivoted readily from appeasing the European appetite for human labor to quenching its thirst for oil. "The kings of the countries where the palm tree grows find that the labour of their subjects, in collecting the fruit and extracting the oil, is far more remunerative to them than the selling of these subjects into slavery," a chemist named Leopold Field told a London audience in 1883. "Being as keenly alive to their own interests as any white men can be, they have become humane as a matter of business. By encouraging the influx of European goods for their native productions, they have brought about their own civilization far more rapidly than could have been effected by the simple spiritual pressure of missionaries unendorsed by self-interest."

Over the years, the leading Delta middlemen had built up corporation-like entities known as "houses" by acquiring people, both slave and free, through a complicated political system involving routine skirmishes with neighbors and rivals. The more "clients" a middleman could claim, the more expansive his geographical domain and the greater his potential to source oil. While the smaller houses numbered in the hundreds of clients, "royal" ones might claim oversight of many thousands. (Middlemen often were chiefs and kings whose power had derived from family lineage, though in rare instances they had risen from the lower rungs of society.) Competition among the houses was fierce, with ongoing clashes leading to the buildup of military fleets: canoes mounted with cannons and large-caliber guns would glide through the creeks alongside those ferrying oil. At other times, the middlemen might conspire to hold back their oil if they felt the prices were too low. "They are a people of great interest and intelligence," noted the anthropologist Percy Amaury Talbot in a self-owning passage from his 1932 book *Tribes of the Niger Delta*, "hard-headed, keen-witted, and born traders. Indeed, one of the principal agents here, a [European] of world-wide experience, stated that, in his opinion, the [Delta traders] could compete on equal terms with Jew or Chinaman."

Things had begun to change around the middle of the century, when the introduction of quinine had made it possible for Europeans to begin penetrating the malarial interior. At the same time, the arrival of steamships on the continent meant that traders could, during the rainy season, travel hundreds of miles up the main course of the Niger River. By the 1860s, newly arrived traders had pushed as far north as the towns of Aboh and Onitsha, located in the palm oil–producing regions, constructing crude storage and processing facilities known as "factories" on the riverbanks along the way. When their vessels tried to steam back downriver laden with fruit and oil, the traders often would come under attack by locals allied with the coastal middlemen. (Annoyed at losing out on the relationships they'd built up over decades, the Liverpudlians were happy to provide their native allies with arms and ammunition.) Swamp-bound as they were, the middlemen could access little more than fish and relied on trade with the interior for feeding their populations. In response to the ongoing assaults, including, at one point, the sinking of a Holland Jacques steamship, the British consul—the first having been appointed to the region in 1849—would send gunboats upriver, burning villages along the way to remind the natives who was in charge.

As tensions rose, the traders increasingly turned to London for help. The Invisible Empire was all well and good, insofar as a handful of consuls and gunboats could keep business interests in line at little cost to the taxpayer back home, but by now the stakes had risen considerably. In an 1871 memo to the Foreign Office, Charles Livingstone, the brother of David and then serving as consul for the Bights of Benin and Biafra (having succeeded Burton), noted that no fewer than twenty palm oil companies were trading in the region, having established sixty factories on seven rivers. "Their property in the Rivers in English manufactured goods, Hulks, Houses, River Steamers, Schooners, Lighters is not much short of a million pounds sterling," he wrote, "and is often estimated at a higher figure." Yet the system for exporting the oil—some thirty thousand tons of it annually—remained far from ideal. "In all the rivers, our agents trade only with the tribes that own the river-mouth, and are not oil-producers, but merely oil-brokers or middle-men," he continued. "Gladly would the nearest oil-producers come and trade direct with the whites, to the advantage of both, and most willingly would some white and many black traders go up to the markets of the oil producers, but the black brokers are strict protectionists, and allow no trade with white or black, except what passes through their own hands, at their own price. . . . All are protectionists, and have ever been. Savage Africa is the home of protection and heathen."

An avowed atheist, Goldie had no quarrel with heathens, but he *was* bothered by the protectionism, namely because it wasn't his own. As soon as he got back to London—there would be no settling in the unwholesome miasma for this young aristocrat—he got to work on his competitors, and within two years he'd convinced Holland Jacques' three main rivals to merge with him. By 1879, Goldie, who'd quickly made a name for himself in the palm oil world—as much for his bombast, it must be said, as for his brilliance—had announced the formation of the United African Company, or UAC.

With the help of some fifteen hundred new recruits, he began send-ing steamboats up to the palm groves to exchange firearms, gin, and rum for oil directly with the region's chiefs and kings—middlemen be damned. The UAC built dozens of factories along both the Niger and Benue Rivers (the latter flowing into the former from the east) and was soon competing with the native houses for economic and political control. The king of Brass, located at the central mouth of the Niger and once among the primary figures of the palm oil trade, found his business so under siege that he appealed directly to the British Foreign Office:

> Many years ago we used to make our living by selling slaves to Europeans, which was stopped by your government, and a treaty made between you and our country that we should discontinue doing so; and that we should enter into legitimate trade . . . this we did and our trade gradually increased until we shipped an average of about 4,500 or 5,000 tons of palm oil per annum. To do this we had to open up places on the Niger, trading stations, or markets as we call them, up as far as a place called Onit-sha. . . . Some years ago the white men began trading in the Niger . . . they did us no harm as long as they went up a long way further than we could go in their steamers. . . . But lately, within the last six years, they have been putting up trading sta-tions at our places . . . (now) we do not send 1,500 per annum. . . . It is very hard on us, in all the other rivers . . . the markets are secured to them, and why should a difference be made for this my river. We have no land where we can grow plantains or yams. If we cannot trade we must starve.

Goldie was unmoved. Both he and the consul at the time, a young striver named Edward Hyde Hewett ("a convivial fellow," writes Thomas Pakenham in his 1991 book *The Scramble for Africa*, "except when he was prostrate with fever, as he was much of the time"), made the case that the only way for Britain to safeguard her growing interests in the region would be through a formal political arrangement. But the officials back in London were having none of it. "The coast is pestilential," huffed the colonial secretary, Lord Kimberley, in an 1882 letter to Prime Minister W.E. Gladstone, "the natives numerous and unmanageable. The result of a British occupation would be almost certainly wars with the natives [and] heavy demands upon the British taxpayer."

Events across the African continent were moving fast, though, not just on the Niger but farther west along the coast and up in Egypt, where the French and other European powers were aggressively jockeying for position. Goldie bode his time, purchasing the assets of the UAC later that year and forming a new outfit, the National African Company (NAC), for which he handpicked each director based on his political connections. By January of 1883, they were knocking on the door of the Foreign Office themselves. The NAC "had no wish for 'exclusive advantages'" to monopolize the trade of the Niger, Goldie assured the assembled suits, but with the French and Germans encroaching on the region's lucrative trade—the Scramble being now fully on—somebody needed to do something to hold them back. If the British were to establish an official relationship with the natives, he argued, whether in the form of a colony or a protectorate, Goldie would be in a position, under a free-trade arrangement, to pull off a legal monopoly. Were the French to gain control, on the other hand, the notorious protectionists would undoubtedly introduce tariffs and block the British from the markets.

What Goldie really wanted was a royal charter, so as to control the entire region himself. "My dream as a child," he would later confess, "was to colour the map red."

No matter that the map was already deeply shaded, a kaleidoscope of hues representing innumerable tribes and kingdoms with long-established hierarchies and social mores of their own. While Goldie had been cocooned in his desert idyll, for instance, a young prodigy named Jaja had vaulted himself to the very top of the Delta's palm oil establishment. Tall and charismatic, Jaja had been sold into slavery as a child but had spent his teenage years quietly observing and noting down every aspect of the industry. By 1863, he'd been named head of a prominent Bonny trading house. In a dispatch from the following August, then-Consul Burton had warned his countrymen about "one Jaja, son of an unknown bushman, a common Negro. He is young, healthy, and powerful, not less ambitious, energetic, and decided. He is the most influential man and the greatest trader in the River, and 50,000 pounds, it is said, may annually pass through his hands. He lives much with Europeans, and he rides rough shod over young hands coming into Bonny. In a short time he will either be shot or he will beat down all his rivals."

In 1866, a great fire swept through Bonny, taking all of Jaja's weapons with it and offering an opening to the young leader's competitors, who quickly hit the rivers in their war canoes. "Gentlemen," came a message from Jaja a few days later, "I beg to inform [the] Court that I cannot fight any more, because I have no house and no carriage and have [no] guns to fight as all were burnt."

Just eight weeks after that, though, now-Consul Livingstone returned to Bonny to find the chiefs unsure as to whether they'd beaten the guy or not. Jaja had moved himself further east, to a place called Andoni, and was now controlling the creeks that led to some of the best palm oil markets in the region. A year later, he broke away entirely and established his own kingdom, calling it Opobo.

King Jaja, once the greatest palm oil trader in the Niger Delta.

He proceeded to occupy strategic points along the rivers and to bar entry to Europeans and African middlemen alike. By 1870, Jaja was selling eight thousand tons of palm oil directly to the British and had become known as "the greatest African living in the east of modern Nigeria."

"Jaja's blockade entirely cuts off the trade of seven English firms which can never go to Opobo," Livingstone fulminated that same year, by which time the Liverpool trade had fallen off by fully half. The new king had rendered bankrupt most of the English firms operating in Bonny and at one point had sent fifty armed canoes to confront a group trying to trade directly with Europeans, murdering several hundred of them in the process. (His rivalry with the European traders notably did not preclude Jaja's sending his children off for a posh Glasgow education or enlisting whites to staff the secular school that he'd built at home. He also made it abundantly clear that no missionary was welcome anywhere near Opobo.)

As the Foreign Office back in London bickered with the Colonial Office over what to do about the Delta—and who should pay for it—tensions on the ground continued to rise. In 1882, the Brassmen, whose situation had only grown more desperate since the king's written appeal, shot at the NAC's boats and attacked several of its factories. Soon after, an upriver tribe looted an NAC factory after killing its manager and four of his staff. Consul Hewett had sent up his gunboats for the usual shelling and village-burning, and was making an increasingly impassioned case for the necessity of a formal political arrangement in the Oil Rivers.

London finally acquiesced, agreeing to the establishment of a "paper protectorate" and closing down a handful of other consulates, including one in Honolulu, in order to pay for the vice-consuls that the administration of the new territory would require. In 1884,

Hewett set out in one of his gunboats from Benin, in the far west of the Delta, and stopped in nearly every city-state from Warri to Andoni, plying their chiefs and kings with rum and coercing them into signing agreements that ceded their political rights to Britain. Though Jaja agreed to put his name to a "protection" treaty with the Crown, he did so only with the provision that it not include an article ensuring Britain "free trade," well aware that doing so would open his hinterland markets to the outsiders.

Later that year, with conflict raging across the African continent, German chancellor Otto von Bismarck convened a conference of European diplomats in Berlin. Its aim was to hash out the endless territorial claims and counter-claims in hopes of averting all-out war. Goldie, having by then bought out even his French rivals, went along with the British delegation to advise behind the scenes, furnishing them with Hewett's signed treaties as evidence of British dominance in the Delta. The General Act of the West Africa Conference of 1884–85 divvied up the continent like a birthday cake (for "furthering the moral and material wellbeing of the native populations . . . and bringing home to them the blessings of civilization"), with the prize that was the palm oil–endowed Delta, now officially the Oil Rivers Protectorate, awarded to Great Britain.

In July of 1886, as reward for having all but single-handedly secured a vast swathe of the continent for the Crown, Goldie was finally given the charter he'd so long desired. It empowered the company "to administer, make treaties, levy customs and trade in all territories in the basin of the Niger and its effluents." The board of directors in London became the "Council" of the newly named Royal Niger Company, or RNC, with Goldie taking on titles as deputy governor and political administrator. Company agents fanned out across the rivers, signing hundreds more treaties, while Goldie put in place laws requiring that any "vessel, boat, canoe, or other craft" coming from outside the Niger Territories enter and clear at the coastal city-state of Akassa, site of the RNC's main compound, to be certified by customs authorities. He mandated that traders pay duties on tobacco, salt, guns, powder, gin, and rum—basically anything they were likely to be bringing in—and, on their way back out, to pony up for taxes on palm oil and kernels.

Next, Goldie invested in a fleet of custom-designed gunboats capable of navigating the Niger year-round (for "pacifying" the "robber chiefs . . . who were ever-ready to plunder our factories") and established the Royal Niger Constabulary, a lavishly equipped army and navy consisting of British officers and some 150 Nigerians that, over the next fourteen years, would engage in more than one hundred military expeditions. His agents continued to pile up treaties ceding to the RNC "the whole of the territories of the signatories" and conferring the right to exclude anyone attempting to do business in "company" territory. (So much for the protectionist French.) Natives throughout the Oil Rivers and up the Niger now were obliged to trade only at RNC posts, and if they attempted to sell to others or to conspire against the company in any way, the steamers would chug up the murky water and fire into villages, setting huts ablaze and sending women and children fleeing into the forest as the men tried to hold their ground with arrows, spears, and antiquated muskets.

Jaja, meanwhile, had been amassing ever more power in the east. By 1887, the new consul, Harry Johnston, was informing the young king that his direct shipments of duty-free oil to Britain constituted unfair competition and ordering him to stop levying duties on English traders. When Jaja refused to comply, Johnston set out for Opobo and invited the upstart onboard his gunboat for a meeting. Jaja refused.

"I hereby assure you that whether you accept or reject my proposals tomorrow no restraint whatever will be put on you," Johnston insisted in a note. Finally convinced, the Nigerian climbed onboard and was promptly informed that he could either face trial for obstructing trade or expect a British bombardment. He was hustled off for a trial at Accra, in nearby Ghana, found guilty, and immediately sent into exile in the West Indies. On August 9, 1891, his obituary appeared in the *New York Times*, under the headline "King Ja Ja of Opobo Dead: England's affair with him and the end of it." Visitors to Opobo today will find a statue of the young king, still considered among the greatest Nigerians of the last century, standing in the center of town, not far from the three-story house that the trailblazing monarch had imported from Liverpool.

By 1892, RNC stations dotted the lower Niger, with the company controlling its every creek and tributary. Goldie had, in defiance of his charter, locked out all competition, whether European or African. Even the Liverpudlians, once staunch allies of the middlemen, were by then collaborating with their crafty compatriot. The once-prosperous Brass had been reduced to sneaking through the creeks at night to smuggle oil, only to be fired upon by Goldie's men, who also seized whatever yams and cassava the tribesmen were shuttling from upcountry in hopes of keeping their people alive. The end of 1894 saw the Brass starving and a smallpox epidemic tearing through their villages. RNC employees, meanwhile, had taken to behaving like unruly frat boys, offending local customs at every turn. The rape of a local Brass woman by a company clerk would prove the very last straw.

On the night of January 28, 1895, a king named Koko, who'd recently replaced the late, letter-writing head of the Brass, led his desperate countrymen on an outing to finally seek justice. Throwing off his Western trousers and buttoned shirt for his traditional "war dress," as Achebe would have called it, he covered his body with chalk and draped a string of monkey skulls around his waist. He and one thousand tribesmen then sprinkled themselves with holy water and set out from Nembe, the capital of the city-state, in a fleet of canoes. Paddling into Akassa under a shimmering moon, they proceeded to loot and wreck every one of the RNC's workshops, customs houses, wharves, engineering shops, and offices. They also snuck into the huts of the sleeping African staff and slaughtered seventy-five of them—most of the Europeans having escaped on a launch—then loaded prisoners and loot into their canoes and headed back to Nembe. There, on the orders of Koko, the Brassmen proceeded to murder their captives before cooking and eating their body parts in a ritual sacrifice intended to quell the rampant smallpox.

"Our boys fired, killed and plundered, and even the innocent provisions sellers were captured and killed likewise," Koko would acknowledge soon after the incident, in a letter to the British Foreign Office. "If the Queen of England was acting in like manner as the Niger Company the whole of Africa would have been dead through starvation . . . the Company is not the Queen's man; and instead if we Brass people die through hunger we had rather go to them and die on their sords [sic]."

Back in London, Goldie received the news with astonishment. "We always looked on Akassa as being as safe as Picadilly," he said, before vowing to retaliate. His first step was to persuade the Foreign Office to telegraph orders to Claude MacDonald, consul-general of the Niger Coast Protectorate (the Oil Rivers Protectorate having been renamed in 1893), demanding that the Brass give up their weapons.

"People of Brass understand thoroughly reparation has been and is being made for past offences," a defiant MacDonald telegrammed back. "Canons, canoes, plunder and prisoners have been surrendered, and the chiefs who took part in the atrocities fined; towns are destroyed, trade almost ruined, women and children starving in the bush; hundreds have been killed; smallpox has been raging; the rainy season is beginning. I have seen all this and visited the towns destroyed," he concluded. "I most strongly deprecate further punishment in the name of humanity, and request a settlement of the question."

The chastened colonial secretary told MacDonald to instead lift Goldie's blockade, following that order with another for an inquiry into the conduct of the RNC. When rumors began circulating among the Brass that the man selected to lead the investigation would recommend that their district be transferred to Goldie's company, Koko and his chiefs drafted a letter to the Prince of Wales begging not "to be *oppressed* and *exterminated* in *revenge*." It was their "*grievances* and *sufferances* here under the Royal Niger Company," they wrote, "which have driven us to take the laws into our own hands by way of revenge in looting the Company's factories at Akassa and attacking its officials for which we are now *very very sorry* indeed, *particularly* in the *killing* and *eating* of parts of its employees."

In the wake of the Akassa attack, Goldie published a piece in the *London Times* in which he boasted of having helped jettison Jaja and faulted Consul-General MacDonald for having defended the Brassmen. "Jaja was deported . . . because he was a big monopolist," MacDonald fired back in a letter to the Foreign Office. "Now we have wiped the floor with the Brassmen because they have endeavoured to go for the biggest monopolist of the crowd—the Royal Niger Company. As I daresay you are aware, in the vast territories of the Niger Company there is not one single outside trader, black, white, green or yellow. The markets are all theirs. They can open and shut any given market at will, which means subsistence or starvation to the native inhabitants of the place. They can offer any price they like to the Producers, and the latter must either take it or starve. And why, in heaven's name, why? Because [the Company] must pay their 6 or 7 per cent to shareholders."

Plus ça change, as they say. Goldie, who by that point had dropped the Germanic-sounding "Taubman" from his name, would go on to have one last hurrah in the Delta, when, at the end of 1896, aged fifty, he traveled down to lead company troops in battle against a pair of Islamic warrior states north of the Niger Delta. The rulers of Nupe and Ilorin had long been challenging British authority on the middle of the river, and now the French were making noise about the illegitimacy of Goldie's sprawling commercial sphere. The area claimed by the RNC, they said—some three hundred square miles—covered a much wider swathe than the territory over which the arrogant Brit actually held treaties. Now they'd begun pushing south with an aim toward claiming some of it for themselves. Whoever seized that land would be in a position to press northeast to the rich states there, and on to the coveted Sudan.

In January of 1897, with a force of some 30 Europeans, 513 African soldiers, and 900 local porters, and armed with the latest in battle-front technology, Goldie marched from the company's base at Loko-ja and defeated a Nupe army numbering in the tens of thousands. Three weeks later, he moved on to Ilorin, shelling the city from close range and watching as it went up in flames. A year after that, the Niger Convention was signed in Paris, ceding to the British significant territory both northwest and northeast of the Delta.

By that time, though, even Goldie's countrymen had begun to tire of his antics. Liverpool newspapers were running headlines accusing the RNC of "murdering natives" and drowning its territories in gin. The colonial secretary had had enough of the endless protests against Goldie's illegal monopoly, of the complaints from the Liverpool merchants about his failure to effectively govern, and of the ongoing entreaties from the desperate Brassmen. On January 1, 1900, the British government finally revoked the Royal Niger Company's charter and took control of its possessions, paying Goldie and his partners £865,000—some $140 million in today's currency. His territory and the adjacent land were reorganized into the protectorates of Northern and Southern Nigeria, precursors to what today makes up Africa's largest economy.

Goldie may not have colored the entire map red, but he did manage over the years to dye a significant portion of it. The effort to protect Britain's palm oil interests had served as prelude to the establishment of its rule in the Delta, and the one-percenter's private empire would become the model for chartered companies on the continent moving forward. Goldie is still referred to as the "Founder of Nigeria" and is routinely compared with Cecil Rhodes, the mining magnate credited with having claimed much of southern Africa for the Crown. Goldie died in a London hotel room in 1925, at the age of seventy-nine, but not before having been knighted by Queen Victoria and having a rare West African venomous snake, *Pseudohaje goldii*, named in his honor. He destroyed all of his papers and forbade his children from ever writing about his life. In any case, George Dashwood Goldie wouldn't be the last outsized Brit to seize upon the unsuspecting oil palm fruit as a way of leaving a lasting stain on the African continent.

2

The Flavor of Home

*It is as if one had taken a cutting of Africa
and rooted it in Brazilian soil, where it bloomed again.*

—Roger Bastide, *The African Religions of Brazil*

WE SAT low in the open-air craft, shouting against the roar of the
motor as our hair flapped wildly in the breeze. In the minutes before
pushing off from the dock, we'd been joined by ten or so locals,
who'd tossed their backpacks and plastic bags of fruit into the prow,
then claimed seats amid the jumble of jerry cans and other sundry
rural-commuters' effects. Now we were gliding over the surface of
the river, bracketed on either side by a two-toned wall of mangroves,
their dark roots giving way to skinny white trunks that erupted into
scraggly mops against the sky. It was "as if they had lost all count of
the vegetable proprieties," remarked Captain Frederick Lugard of a
similar view from a Niger Company steamboat circa 1890, "and were
standing on stilts with their branches tucked up out of the wet, leaving their gaunt roots exposed in midair."

We might have been in the Niger Delta ourselves. In fact, we were an ocean away, almost directly across the Atlantic from Goldie's old dominion in the Oil Rivers. Along with a few acquaintances from Salvador, the capital of Brazil's Bahia state, I had driven some sixty miles south to reach this coastal maze of tiny islands and riverine canals. Soon after arriving in Brazil back in 1500, the Portuguese had set about enlisting this region's indigenous Amerindians to toil on their nascent sugar and tobacco plantations. When the natives proved too few for the growing enterprises—the diseases brought by the newcomers hadn't helped the situation—the Portuguese took to importing labor from abroad. By 1888, when slavery was finally abolished in Brazil, some 5 million enchained Africans had been

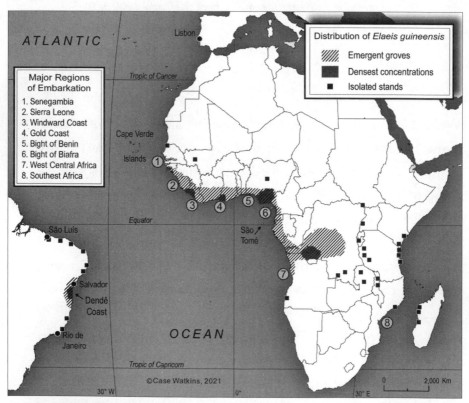

Distribution of *Elaeis guineensis* in Africa and Brazil with major regions of transatlantic slave embarkation, 1500–1900.

marched ashore in Brazil—nearly half of the 10.7 million slaves that disembarked in the Americas during the entire course of the trans-Atlantic trade.

The regions from which those millions of men, women, and children were abducted, extending from today's Senegal in the north down to Angola, correspond almost exactly with Africa's sub-spontaneous oil-palm groves. (Some 54 percent of the Africans who reached Bahia through the trade came from ports of embarkation in the Bight of Benin.) Having made the journey with the captives, the fruit would become a potent symbol of all that they'd been forced to leave behind, a phenomenon that has been documented in fascinating detail by Case Watkins, a geographer at James Madison University.

Palm oil would have numbered among the staples in the African ports frequented by the Portuguese and other traders during the early days of colonialism; the 1642 ledger of a Dutch merchant named Pieter Mortamer documented an ongoing exchange of the oil and kernels moving from Angola and the island of São Tomé, off the coast of Gabon, to Brazil. On the ships crossing the Atlantic, women captives often were charged with pounding and boiling the fruits to produce oil for seasoning the gruel fed to their fellow slaves. When British forces seized a Spanish schooner in 1838, for instance, they reported that, "though no slaves were actually found onboard," there were "all the circumstances which attend the fact of her very recently [having] landed a cargo of slaves"—among them "three slave-boilers . . . called 'three cauldrons for clarifying palm-oil.'" As the slave ships neared port, palm oil would have been slathered into the captives' skin with an aim toward hiding their wounds and scars and enhancing their attractiveness to buyers.

The kernels eventually were planted in the Bahian soil, with oil-palm groves centered in this coastal area and extending south some three hundred miles from the colonial capital. The region remains the only place in the New World where you'll find dense natural stands of oil palm just like the ones in Africa. In 1699, a British privateer named William Dampier wrote about a local population of enslaved Afro-descendants in this part of Brazil "so numerous, that they make up the greatest part or bulk of the inhabitants," and described "palm berries (called here dendees)" that "grow plentifully about Bahia" and the largest of which are as big as walnuts. "These are the same kind of berries or nuts as those they make the palm-oil with on the coast of guinea, where they abound: and I was told that they make oil with them here also."

The word *dendê*, as the fruit is still known in Brazil—palm oil is technically *azeite de dendê*—derives from the Kimbundu term *ndende*. Most of the slaves sent to the country during this period would have spoken one of several Bantu languages native to West and Central Africa, with northern Angola's Kimbundu serving as a sort of lingua franca among them. In 1751, a Portuguese noble known as the Count of Atouguia, having traveled to Bahia to evaluate the commercial potential of various crops, wrote of an oil "for the blacks to eat . . . from a seed, here called *dendê*, that is so abundant its price rarely increases to that of olive oil from Portugal."

As the count's report suggests, the colonial masters paid little attention to *dendê*. Their concern was with cash crops, and tobacco and sugar, unlike oil palm, didn't take well to the salty coastal soil. To their mind, the land was better used for growing manioc, which could be fed to the enslaved workers who by then made up more than half the population. Beginning in 1639, royal decrees had ordered farmers on the southern coast to plant the starchy tuber, otherwise known as cassava and native to the Americas. When Afro-Brazilians cut down tracts of forest to plant the crop, they would often, as their

African ancestors had done when cultivating yams, spare some of the oil palms. As early as the eighteenth century, many slaves who achieved freedom—however conditional it was—continued to farm, usually focusing on manioc fields interspersed with oil palms. *Pirao*, a porridge of manioc flour seasoned with palm oil, remains a pillar of Bahian cuisine, and is often served with *moqueca*, a spicy, palm oil–rich stew built around the fish, shrimp, and crab abundant in these mangroves.

Traveling through Bahia in 1819, the German naturalists Johann von Spix and Karl von Martius also noted the prevalence of *dendê*. "The preparation of palm oil is done by slaves, and because of this, without great care. . . . They use this common oil in cooking, where it is greatly appreciated, especially by Blacks; and also in lamps and as an ointment. They consider the ointment a treatment for skin diseases. . . . Frequently one observes, on the streets of Bahia, a black man rubbing himself with cooked *dendês*, preparing himself, that is to say, for the nocturnal dances (Candomblé)."

I'd traveled down to coastal Bahia to see for myself how *dendê* had come to dominate not just the landscape of this place but also its culture, including in those mysterious-sounding dances. A Salvador-based chef named Alicio Charoth had arranged with a local *terreiro*, or religious temple, to cook a celebratory meal for its members, and he'd invited me to come along. Fit in a black tank top and with close-cropped salt-and-pepper hair, the fifty-eight-year-old ranks among his country's more thoughtful culinary celebrities. Having left behind the restaurant scene in Europe, he has, for the last several years, been focused on unearthing and celebrating the indigenous cooking of his home. With the meal at the *terreiro*, he intended to pay tribute to the ancestors who'd preserved the foodways of this region over generations. Many of the folks with whom we'd be spending the day—we were in a place called Cajaiba, south of the city of Valença—had descended from residents of the region's *quilombos*, as the settlements founded by runaway, escaped, or abandoned slaves are known. It was in such communities that *candomblé* had thrived. The complex system of beliefs—it goes well beyond "nocturnal dances"—has its roots in practices brought by the Yoruba and other West African peoples and incorporates elements from both Roman Catholic and Native American traditions. (Because *candomblé* was long condemned by the Portuguese authorities, its adherents often concealed its practices and symbols inside those of the dominant religion.) So prominently does palm oil figure into its traditions that devotees of *candomblé* are known not just as *povo do santo*, or people of the saints, but as *povo do dendê*.

Throughout the nineteenth century, the ships trafficking in slaves often would carry, in addition to palm oil, kola nuts, cowry shells, and textiles—items that had been snuck onboard for eventual sale to Africans and their descendants in the New World. In a letter published in 1802, a Portuguese bureaucrat named Luis dos Santos Vilhena wrote that "black women" were collaborating with European wholesalers to sell cloth, "most of it either contraband, pilfered, or purchased from foreign vessels, . . . from trading posts along the coasts of Guinea and Mina, thus avoiding the customs duties which [should] go to His Majesty." These women also, he complained, "sell on the streets the most vile and insignificant foods, . . . *carurus*, *vatapás*, porridges, . . . *acarajé*, *ubobó*," all of them centered on palm oil.

By the late 1800s, the commerce between Bahia and Africa having dried up with the end of the slave trade, domestic *dendê* cultivators had moved to fill the vacuum. In 1878, the Salvadoran daily newspaper *O Monitor* reported the arrival by boat of twenty barrels of palm oil from Brazil's southern coast. Nine years after slavery's official abolishment, officers in Bahia inventoried the contents of a home owned by a recently deceased former slave and found a mortar and pestle, a large jug, and a small bottle of palm oil, suggesting that she'd been processing the oil for sale. The country's elites, meanwhile, continued to demean the staple and the culture in which it figured. "Naturally, as the number of proselytes of this [*candomblé*] fetishism continues to grow," read a 1905 article in another Salvadoran paper, the *Diário de Notícias*, "so too will the number of houses of worship and their clerics—which is to say, those who would exploit the credulity of fools. While harmful to society, it enriches traders of the so-called *azeite de dendê*. . . . For this mob of ignorant cretins, there is no evil, moral or physical, that cannot be vanquished by certain herbs mixed with a little *azeite*."

In the century since, the "cretins'" vile and insignificant foods have made their way to the proud forefront of Bahian culture. Among the top attractions for visitors to Salvador today are the city's *baianas*, local women decked out in colorful head wraps, hoop skirts, and lace-trimmed blouses—costumes derived from traditional African dress—and serving up those very dishes from makeshift stalls around the city. Primary among their offerings are the aforementioned *acarajé*, spicy, deep-fried black-eyed-pea cakes like the ones the women's ancestors would have fashioned over open fires in Africa. (Those, in turn, having descended from falafel crafted with chickpeas brought south centuries earlier by Arab traders.) The word *acarajé* derives from the Yoruba *akará*, or "fire," combined with *ajume*, "to eat." The dish, often slathered with *vatapá*, a paste made from dried shrimp, coconut milk, palm oil, and peanuts, and topped with prawns, is considered a sacred offering in *candomblé*. On the "street

On the streets of Salvador, frying up falafel-like *acarajé* in palm oil.

of *acarajé*," a narrow alley tucked into Salvador's sprawling São Joaquim market, shelves display bottle after bottle of deep-orange *dendê*
alongside cloudy coconut milk and trays mounded with tiny brick-
red prawns, white *fradinho* beans, and inch-long *malagueta* chiles.

• • •

Charoth had a far more elaborate meal in mind. On the boat ride over, he'd been deep in conversation with a local woman he called Dona Maria about the coast's natural wealth and how it figures into the region's spiritual traditions. Over decades—she is now in her eighties—Maria has stood at the center of this *quilombo* community. (Though the current *terreiro* dates only to 1995, its roots extend back more than a century.) Entering the compound from the road, the two stopped every few feet to pluck leaves from bushes and yank unfamiliar-to-me fruits from trees, tossing around ideas about how the various finds might figure into the meal. Pulling open a bristly, golf ball–sized pod to reveal a quarry of tiny red seeds, Maria explained how her dad would fashion such *urucum* into a varnish that he brushed on furniture to make it shine. (Known in the United States as annatto or *achiote*, the plant is prized among natives of the Brazilian Amazon as the basis for some of their body paints.) A few hours later, Charoth would pound the seeds into a paste that he used to shellac fillets of *tainha*, a local fish that he'd bargained for on the road coming in.

Maria pulled off a few foot-wide leaves sprouting from thick stems, like Swiss chard from the dinosaur age, and presented them to Charoth. He would eventually chop the greens and sauté them in *dendê*, ending with a dish nearly identical to one that I'd been served in rural Liberia. Among the other elements he'd incorporate into his feast were a bunch of *licuri*, grape-sized fruits with brown skins from a palm native to a region farther inland, and pear-like fruits called *jambo vermelho*, both of which we'd bought along the roadside driving down from Salvador. Also on prominent display during that trip: a series of black-and-white signs hailing our arrival to the "Costa do Dendê," or Dendê Coast. After centuries of basically denying the plant's very existence, the Brazilian government had gotten wise to the tourism potential of the oil palm–rich region, and had officially designated this seventy-mile strip as such in 1991.

We arrived at the *terreiro* a little after 8 a.m., when the community was just rousing itself from the previous night of partying. Charoth had planned our visit to coincide with something called the Festival of Iemanjá, an early-February celebration honoring the goddess of the rivers, and we'd shown up some twelve hours into the reveling. A pair of squealing boys chased after a flock of chicks, while the adults—most of them women, and all dressed in white, from strapless sundresses and long skirts to pajama-like pantaloons—wandered groggily about. Some, en route to the bath, bore metal pans of water. More white dresses, shirts, and pantaloons hung drying in the morning sun. The festival, which is celebrated throughout the region, involves the adherents of *candomblé* dressing in white and offering floral tributes to the goddess at the water's edge. Towering overhead, the canopies of oil palms mingled with those of coconut, papaya, and other trees. The lower trunks of many were beribboned in strips of white cotton.

A large woman with skin the color of milky tea emerged from a building and engulfed Charoth in a bear hug. Wearing a white eyelet dress and matching headscarf, Mãe Bárbara, high priestess of the *terreiro*, proceeded to walk us around the compound, which comprised a dozen or so single-story cement structures, painted white and featuring wooden doors and corrugated-metal roofs. She pointed out the small museum and school, and I noticed above one door the stenciled words "Ancestry, Memory, Education and Resistance." All of the community's activities—from cultivating crops and hosting educational workshops to cooking and participating in various dances—figure into the central *terreiro* mission that is "world-building," the act of reclaiming both identity and power.

Dressed in cargo shorts and flip-flops, and, in proper Brazilian fashion, drinking and smoking his way through the job, Charoth threw himself into the meal preparation. He laid out a series of oil-palm fronds and banana leaves on a rickety table in the yard to serve as a sort of tablecloth, and, shuttling from there to the open-air kitchen, a crude hut crafted from bamboo, began to improvise with the ingredients that he'd accumulated. An assistant who'd traveled with us from Salvador seasoned the *aratu* crabs we'd bought at the dock with lime, then cracked open a coconut using a sharp stone. Charoth would eventually fry small balls of the crab mixture in *dendê*, resulting in fritters that he called his personal take on *acarajé*. "*Dendê* is more than just an ingredient," he told me. "It's symbolic of the Bahian people and our culinary tradition, an example of resistance, and a break with those who colonized us." Dona Maria turned up at the kitchen a few hours later, bearing a plastic bag of manioc flour and yet another liter bottle of *dendê*, sourced from an artisanal mill down the road. "Even after cooking for so many years," said Charoth, sautéing the fish in the oil, "what interests me most is not what's on the plate itself but rather what spills over."

Like Alicio Charoth, the Salvador-based artist Ayrson Heráclito relies on *dendê* to convey something essential about the Bahian character. The son of an African-indigenous father and a Portuguese-Italian mom, Heráclito became an initiate of *candomblé* two decades ago, and he has since incorporated its symbolism into nearly all of his work. His arresting *Divisor* ("Divider"), a Rothko-esque installation in which a band of viscous red-orange palm oil floats suspended in

Ayrson Heráclito's *Divisor*, incorporating salt water and palm oil.

a glass tank over a clear expanse of salt water, acts as a meditation on the legacies of slavery and the trans-Atlantic journey known as the Middle Passage. The curators of a recent exhibit about Bahia at the Fowler Museum, in Los Angeles, chose the installation to be the first piece visitors saw upon entering the show. Over the course of the slave trade, Heráclito told me, the Europeans made it a point to separate people who shared common languages and customs. The *dendê* in *Divisor* doesn't just represent the African blood spilled at the hands of slave traders, but also "a connecting space" linking Afro-Brazilians with their ancestors across the sea. "The Atlantic is the uterus of Black America," he said, the place where "all of the Africans got mixed." Inside the glass tank, the salt water and the palm oil coexist, he added, "but they never actually blend. It's a form of thinking of interracial relations."

Heráclito, who frequently exhibits in such African cultural centers as Bamako, Dakar, and Kinshasa, has also performed ritual cleansings at the House of Slaves on Senegal's Gorée Island, where captives underwent processing before being shipped overseas, and at a former plantation site here in Bahia. Incorporating the traditions, symbols, and foods of *candomblé*, the ceremonies serve as a sort of "liberation" from the pain of slavery, he said, and a way to heal deep historical wounds.

Back at the *terreiro*, such preoccupations were on prominent display. Small shrines incorporating metal, straw, and other materials hung above the doorframes of many of the outbuildings, along with signs listing the ritual foods and symbols associated with *orishas*, the human-like deities said to intervene in the lives of *candomblé* practitioners. Each *orisha* has a favorite dish and color. Tributes to Oshala, for example, the creator of human life and a symbol of wisdom and peace, include fish, bread, and *dendê*. The oil, which generally represents blood, also is used to mollify warlike deities, and clergy use it to douse the carcasses of animals during the preparation of ceremonial meals. (So central are the *orishas* to Bahian culture that a group of twenty-foot-tall fiberglass versions of them dance on the surface of Salvador's Dique do Tororó, a large lake in the middle of the city.)

A little after noon, I made my way over to the main building, where the community members had begun to assemble. Earlier, they had walked in procession to the nearby river bearing flower arrangements and plates of prepared food balanced on their heads. Iemanjá, the deity at the center of the week's celebration, had originally been a goddess of the Yoruba people, associated in particular with western Nigeria's Ogun River, but in her Brazilian incarnation she presides over the ocean as well as the rivers. Her physical manifestation incorporates elements of both mermaids and the Virgin Mary.

The crowd began to gather in a clearing where Charoth was com-
posing his edible tableau. He had carefully arranged a dozen oil-
palm fruits atop a five-foot expanse of the sautéed greens, tucking
them in amid fish heads, crab fritters, hard-boiled eggs, plantain
chips, poached *jambo*, and other elements of the feast. He'd purposely
declined to provide plates or silverware as a way of encouraging the
community members to eat with their fingers, just as he remembered
doing while visiting his grandparents in the Bahian countryside as
a child. The idea was for everyone to acknowledge the region's rich
heritage, including the unique landscape nurtured over generations
by their African ancestors.

Everyone happily dug in, and it wasn't long after that the proceedings took a supernatural turn. It began when Mãe Bárbara suddenly
emitted a sort of croaking moan, her eyes closed and her head falling
back in the manner of a junkie as the needle goes in. Now clutching
her stomach and rocking back and forth, she looked as though she
might be going into labor. But soon the other women were clasping
their stomachs too, and chanting *"Yuh! Yuh! Yuh!"* as they clapped in
a frenzied call-and-response. "The saint has come," whispered my
local fixer. "They're happy."

The group began to dance-shuffle its way into a queue and one by
one embraced Mãe Bárbara. Eight guys were now pounding on tall
drums, and the crowd swayed and sashayed into a kind of spaced-out
conga line. Watching the scene play out, I was briefly transported
back to western Kenya, where, one early dawn after observing many
hours of a similar sort of ritual dancing and chanting, I'd looked on as
some local Luhya women attending a circumcision ceremony became
similarly "incorporated" by spirits.

"Any study of the colonized world," Frantz Fanon writes in his
1961 book *The Wretched of the Earth*, "must include an understanding

of the phenomena of dance and possession. The colonized's way of relaxing is precisely this muscular orgy during which the most brutal aggressiveness and impulsive violence are channeled, transformed, and spirited away."

I can't pretend to know what sorts of emotions underlay the out-pouring of energy I witnessed that afternoon on the Brazilian coast. But I do know that this image—of a people somehow possessed, whether by fear or anger or frustration or joy—would become a recurring theme over the course of my palm oil reporting.

3

The Napoleon of Soap

This man of vast commercial ideas
has now launched himself into West Africa, into the Congo,
where he can become an enormous force for good or for ill.

—journalist Edmund Morel, describing William Lever

AS THE crow flies, the post-industrial town of Bolton sits just ninety-seven miles from the eastern shore of the Isle of Man, but the modest home into which William Hesketh Lever was born in 1851 existed a world away from the servants' quarters and sweeping lawns of the Nunnery, as the Goldie-family pile was known. Bisected by the River Croal, a stagnant murk said to serve as the public sewer, the fast-growing city then crammed some seventy thousand residents into its jumble of factories, mills, and warehouses. ("Even in the finest weather," wrote the philosopher Friedrich Engels of the town in 1844, it is "a dark, unattractive hole.") Lever, who arrived on the scene just as George Goldie would have been graduating to a primary-schooler's silver spoon, was the seventh of an eventual ten children, and the family's first boy. He was educated at the local parochial school, and by the age of sixteen he had gone to work full-time at his father's wholesale grocery. By twenty-one, he'd risen to partner.

Standing a stout five-foot-five, and with notably small hands and feet, Lever was physically unprepossessing, with little about him, save for the unblinking celadon eyes, that hinted at the single-minded ambition that would come to define his life. He married early, to a local woman named Elizabeth Ellen Hulme, and set immediately about upping the game on the family business. After absorbing another wholesale shop, Lever began making regular trips to Ireland and Holland to buy the butter and eggs then flying off of British grocery-store shelves. Cutting out the middlemen ensured a leg up on the competitors. With an eye ever trained on the shifting commercial landscape, he took note of the rising number of middle-class city-dwellers, and of their growing preoccupation with cleanliness. On visits to the docks in Liverpool, just twenty-five miles from Bolton, he'd watched as sacks of oil-palm kernels and barrels of oil were offloaded down the ramps of ships newly arrived from West Africa, to be conveyed to the soap-works that had sprung up nearby.

In 1884, aged thirty-three, Lever introduced a soap brand of his own, named, after much deliberation, Sunlight. Together with his brother James, he took out a lease on a factory in nearby Warrington and announced the formation of Lever Brothers Ltd., dedicated exclusively to the production and sale of soap. Plenty of others were making similar products at the time, but Lever didn't intend to get lost in the crowd. He experimented until he'd found the right blend of ingredients for the ideal lather and rinse—including some 41.9 percent palm-kernel or coconut oil—and took to selling his bars as single units. (Most grocers at the time sliced off portions from long bars and priced them by the pound.) He began wrapping his soaps in parchment to avoid the sweating then plaguing other brands, and tucked them into colorful cardboard boxes. He also spent lavishly on advertising, emphasizing the quality and purity of Sunlight and introducing slogans and giveaway campaigns, tactics he'd picked up from monitoring marketing trends in the United States. He even began buying the rights to contemporary paintings and shrinking down their reproductions for use as packaging and in ads. The British housewife, in the eyes of William Lever, represented a pocketbook waiting to be pried open. His campaigns appealed strategically to this market, promising to reduce facial lines and offering booklets with tips on washing clothes and otherwise keeping the perfect home. "People already in the soap business could have put rings round me on manufacturing soap," Lever would later say, "but none of them understood how to sell soap, . . . advertising, agencies, etc., and left others to look after the works."

Business was so robust that by 1887 Lever's factory was running at full capacity. With little room to expand, he and James found a marshy expanse of land near Bromborough Pool, an inlet of the Mersey River some thirty miles southwest of Bolton, and purchased fifty-six acres there. Not only was the site blessed with facilities for road, rail, and water transport, it was also situated just beyond the reach of the Liverpool Dock and Harbour Authority and its dues-collecting agents. The brothers christened the spot Port Sunlight and allocated twenty-four acres to a soon-to-be-built Lever Brothers Soap Works factory. The remaining thirty-two would, over the next few years, be transformed into a village for workers and their families. An ardent capitalist and a regular at the local Congregationalist church, Lever was a firm believer in self-discipline and hard work, but he was concerned about the impact that rapid industrialization was having on society. He'd seen too many cities like Bolton spring

William Lever in 1920.

up, grimy places where no man would want to live, much less raise a family. Port Sunlight would represent a different sort of community, with each of its homes fronted by a landscaped yard, and with sports pitches and dance halls in addition to its own school and cottage hospital. His aim in building the village, Lever said, was to "socialise and Christianise business relations, and get back again in the office, factory, and workshop to that close family brotherhood that existed in the good old days of hand labour."

By 1894, Lever had taken his company public (James had by then fallen ill, and would retire in 1895), and within three years, Lever Brothers was producing 2,400 tons of Sunlight soap every week. Lever proceeded to introduce new brands, including such soon-to-be household names as Lifebuoy, Lux Flakes, and Vim, and was largely responsible for the fact that, as the twentieth century dawned, the average British citizen was plowing through some seventeen pounds of soap every year.

A man of his late-Victorian age, Lever increasingly viewed the world through the prism of empire. He began setting up new companies abroad and took over or merged with others, eventually manufacturing many of his products overseas. At home, a hummingbird energy combined with a fear of idleness such that the man never seemed to stand still. Would-be interlocutors often were required to accompany Lever on train trips to other appointments so as to maximize his every working minute. At the far end of the journey, a driver would be waiting to convey the first appointment back home. Lever arranged voyages to North America, Australia, and other new markets to check in on the outposts of what was fast becoming a global empire.

Returning to Port Sunlight, he would launch into speeches about his latest theories on everything from factory paternalism and racial determinism to joint-stock companies and democracy. At the same time, he was beginning to immerse himself in such bourgeois pursuits as art collecting, architecture, and gardening, enlisting experts to school him and teaming up with them on building projects that would become ever more elaborate as his fortune grew. "At this our first interview," reported the landscape architect Thomas Mawson, who met Lever in 1905, "he struck me as a veritable Napoleon in his grasp of all the factors dominating any problem he tackled, in his walk and pose, and in his speech, which contained the concentrated essence of thought. There were all the characteristics which we associate with the 'Little Corporal.'"

My own train pulled into Port Sunlight on a bright June afternoon, having departed from London's Euston Station some three hours earlier. There was no need to summon a taxi; the place is tiny, a storybook tableau of manicured lawns and two-story, half-timbered buildings. I rolled my suitcase onto the sidewalk, making a left by a little teashop on the corner, and was headed toward my hotel—housed in that old cottage hospital—when a dozen or so kids, aged about nine, pedaled placidly by in single file, their iridescent-lime vests mini versions of those worn by the adults anchoring either end of the juvenile convoy. A quaint bank, post office, and bakery lined the street opposite a large lawn, across which a smattering of old men in white caps rolled small, dark balls. Turning into the residential area, I strolled amid a charming mishmash of stone, brick, stucco, and wood-beam homes, each with a neat parcel of lawn and cartoon-perfect flowerboxes. The order, the quiet streets, the hushed spectators on the bowling green—all of it underscored the Mayberry-by-the-Mersey vibe, everything pleasant, safe, and in its place. Only the lanky night manager at the hotel, sporting black Chuck T's and with a scraggly goatee and sullen mumble, suggested anything out of sync with the squeaky-clean facade.

I'd traveled to Lever's original headquarters to spend a few afternoons in the Unilever Archives. As mentioned previously, our Little

Corporal eventually went on to lay the foundation for what is today one of the largest multinational food corporations in the world. Based in London, Unilever makes all manner of personal-care and food products, and it still ranks among the top purchasers of palm oil and kernels globally. The company makes a point of touting its sustainability bona fides; on the homepage of its website, the slogan "Brands with Purpose" flashes across gauzy images of thriving farm fields and multicultural models. Lever House, the main office building at Port Sunlight, stands as a grand testament to the early days, with parquet floors, vaulted ceilings, stained-glass domes, and marble busts perched on pedestals. From the paned windows in the reception area, a visitor can look down long galleries where employees sit before oversized computer screens as the sun streams in from above. The 1895 building has been retrofitted with the latest in workplace technology, in the manner of a mod Kimpton hotel slipped into a Grover Cleveland–era bank. Cozy sitting areas with felt-covered chairs in shades like aubergine and acorn telegraph a Scandinavian, worker-friendly ethos. Unilever flagship products like Persil dish cleaner and Dove soap sit inside glass boxes like royal jewels in a museum, as archival stills from the company's past play across a TV screen with captions like "Being Healthy" and "Being Inclusive."

But in the hours I spent hunched over letters typewritten on onion-skin some one hundred years ago, letters beginning "Our Lord," and addressed to "The Right Honourable The Lord Viscount Lever-hulme"—Lever was made a baronet in 1911, a baron in 1917, and a viscount in 1922, and after his wife's death, in 1917, he combined her name with his own to create Leverhulme—and signed variously by expat commercial agents, Jesuit priests, and African chiefs in Lever's employ, the picture that emerged of Unilever's early days amounted to something decidedly less than sunny.

• • •

In 1914, Lever diversified from soap into margarine, a product whose origins dated to a few decades earlier, when—just before the siege that would ensnare Goldie and Mathilda—France's Napoleon III had put out a call for a cheap butter substitute to feed the country's military and underclasses. In 1869, a chemist named Hippolyte Mège-Mouriès had come up with a recipe for beef tallow churned with milk. (His innovation, in turn, built upon a previous discovery, by another Frenchman, of a fatty acid he'd dubbed *acide margarique*.) Two years later, a Dutch dairy family had obtained Mège-Mouriès' formula and dyed the resulting white "oleomargarine" a convincing butter yellow. The new spread had quickly become a staple in working-class kitchens across Holland.

Just after the turn of the century, a German scientist named Wilhelm Normann had developed a process for transforming liquid oils into solid fats, enabling Europe's margarine-makers to substitute vegetable oils for the animal fats upon which they'd previously relied. Hydrogenation had come along at an ideal time: accessing the cheap tallow that they'd been sourcing from Chicago would become difficult just a few years later, when the city's meatpackers would take to organizing and the resulting wage hikes would send prices soaring. The Dutch firms—a second family operation had since entered the market—turned to palm oil for making their margarines and expanded their businesses to Britain and beyond. In 1908, a soap manufacturer named Joseph Crosfield & Sons, based in Lever's original factory town of Warrington, bought Normann's patent and began making its own margarine. Lever followed suit, prompting Crosfield to sue him for patent infringement. Lever won the case, the first in what would become a long string of mostly successful litigations, and established himself as a businessman with whom you tussled at your peril. (Edwin C. Kayser, a chemist for Crosfield, would go on to take a job at Procter & Gamble, in Cincinnati, Ohio, where he would patent two processes for hydrogenating cottonseed oil, leading to the introduction, in 1911, of that company's now-famous Crisco.)

While Lever made it a point to portray himself as a lowly soap-maker, *modest* wasn't the word that acquaintances of the day would have used to describe him. "At very first glance I would put him down as a rather insignificant little fellow," wrote one civil servant. "That impression lasted for a shorter time than it takes to write it. Charm, tact, decision, power radiated from the man's every word, look, and gesture. I had never met a man who was so obviously a megalomaniac and accustomed to having his own way."

With the Brits now churning out margarine, and the Dutch having entered the soap market, competition for West Africa's palm oil and kernels began to heat up. Lever was determined to get his hands on an exclusive supply, not just because he was tired of the uneven shipments and fluctuating prices, but also because he didn't trust the native producers, whom he considered lazy and lacking in discipline. If he could secure a piece of land in Africa and train the workers properly, he figured, he could streamline both processing and transportation while also improving the quality of the oil. (As mentioned earlier, processing helps halt the development of free fatty acids, which compromise its integrity.) Lever approached the British government about leasing land in Nigeria, by then a protectorate, but was rebuffed by the colonial authorities. They'd only just rid themselves of one George Goldie, after all.

Having also been refused land in the Crown colony of British West Africa (which would later become Sierra Leone) and in the Gold Coast (today's Ghana), Lever would eventually turn his attention to Brussels. While Goldie's men had been piling up treaties in the Niger

Delta, the Welsh journalist and explorer Henry Morton Stanley had been traipsing around the Congo, setting up stations and building a rough track around the rapids to enable his patron, King Leopold II of Belgium, to penetrate the terrain downriver. Between 1879 and 1884, Stanley, with the usual assist from cheap gin, had forced or tricked some four hundred African chiefs into signing contracts, invariably written in languages they didn't understand, that handed over their land to Leopold. At the same Berlin conference where the Brits were awarded the Niger Delta, the fifty-year-old Belgian monarch gained diplomatic recognition of his Congo Free State, comprising an area some seventy-five times larger than his native land and rendering him "proprietor" of the world's only private colony.

Leopold quickly grew rich shipping out elephant tusks, ivory then being coveted for carving into everything from piano keys and billiard balls to chess pieces and fake teeth. After John Boyd Dunlop's introduction of the pneumatic tire set off a bicycle craze in the 1890s, the king expanded his operations to include rubber. Natives had long ago discovered that cutting into *Landolphia owariensis*, a species of liana that grows wild in Central Africa's rainforests, released a milky latex that, once dried, possessed elastic qualities suited to innumerable uses. The market for rubber continued to grow as the automobile industry came online and the numbers of telegraph and telephone wires, all requiring insulation, multiplied.

Tall and imposing—and a first cousin of Queen Victoria—Leopold was admired throughout Europe as a thoughtful and philanthropic leader. An outspoken booster of the Three C's, he would go on about moral uplift, abolishing the slave trade, and advancing science, and he readily welcomed missionaries into his colony. It wasn't until the end of the century, when a young clerk named Edmund Dene ("E. D.") Morel began poking around, that the public had any reason to believe that the situation in Leopold's Congo was anything other than he'd been portraying it as. An employee of the Liverpool-based Elder Dempster shipping line, Morel had traveled to Antwerp to review the account books when he discovered that the goods being sent to the Congo Free State were a fraction of the rubber and ivory coming back—and that whatever cargo *was* going out consisted mainly of guns and chains. Beginning in 1901, Morel published a series of explosive articles about his findings, prompting the British Foreign Office to send its consul, an Irishman named Roger Casement, to investigate.

Leopold had declared a huge expanse of his colony to be a *domaine privé*, in which he was free to collect raw materials and levy taxes at will. Given the lack of currency, though, the only way for Congolese men to pay the required head tax was to work it off. Casement learned that state agents had been traveling into villages to draft men for the arduous task of harvesting latex, a job that involved walking deep into the rainforest, perilous with snakes and leopards, and smearing the liquid all over their bodies so that, once dry, they could peel it off—along with whatever hair and skin might come with it. When the men fled into the bush to avoid being conscripted, the agents would take their wives and children hostage. Captured men were required to produce quotas of latex, and those who failed to deliver often would be killed. Members of Leopold's private army, the nineteen-thousand-strong *force publique*, armed with guns and a hippopotamus-hide whip known as the *chicotte*, tortured hostages, raped women, and slaughtered entire families. The soldiers were required to provide the hands of their victims—whether dead or alive—as proof that they hadn't squandered the monarch's precious bullets. Casement's report sparked the formation, in 1904, of the Congo Reform Association, and by 1908 Leopold had been forced to relinquish control of his colony to the Belgian state. It has since been determined that Leopold's regime effectively murdered some 10 million of the colony's citizens.

Smarting from the PR disaster, officials in Brussels saw in William Lever a chance for reputational redemption. An agent of Leopold's nephew, the newly crowned King Albert, traveled to England to propose to the businessman, now well known for his model garden village, a possible investment in the newly rechristened Belgian Congo. Lever wrote a letter to Morel (who'd since left Elder Dempster but continued to campaign against "red rubber," stained in African blood) in which he denied having any interest in establishing oil-palm plantations in the Congo. In fact, he was already engaged in talks with a colonial minister named Henri Dekeyser to do just that. (Followers of the news in England would have recognized the latter's name; he had recently made the papers for commanding a group of soldiers to cut the feet off a Congolese chief's daughter so that he might procure her thick brass anklets.)

In 1911, Lever signed a contract on concessions totaling 1.8 million acres for cultivating and harvesting oil palm in the Belgian Congo, establishing a new company, the Huileries du Congo Belge, or HCB. The thirty-three-year lease conferred ownership of land within five discrete areas, each of a sixty-kilometer radius and priced at 62 cents per acre. Of the five, the most promising concession was at a place called Lusanga, located at the confluence of two rivers and the site of dense sub-spontaneous oil-palm groves. As with those in Nigeria, the forests here had been nurtured over centuries by the local populations, and the oil and kernels the plants bequeathed figured prominently in the people's cuisine and cultural life. Lever's plan was to gather fruit from these extant groves while establishing plantations and waiting for the new trees to mature. He vowed to treat the native Congolese with respect and to always pay his laborers a fair wage, but Morel, for one, had his doubts. "Lever is a man without education," he wrote upon hearing news of the deal, "and without sentiment, a commercial genius, enormously fabulously rich, probably of good heart but also hard, who sees humanity as a vast engine of

Approximate location of the five circles granted to William Lever for oil-palm cultivation in the DRC (formerly the Belgian Congo).

production without soul or desires or ambitions other than the accumulation of *louis d'or*, authoritarian, rarely contradicted. This man of vast commercial ideas has now launched himself into West Africa, into the Congo, where he can become an enormous force for good or for ill."

Lever wasted no time getting his enterprise under way. By the end of the year, he'd shipped one thousand tons of machinery from Liverpool to equip a mill in Lusanga, which he'd promptly rechristened Leverville. By March of 1912, the first consignment of Congolese oil had arrived in Antwerp. (A month later, a single bar of soap crafted from the oil would be placed in a carved ivory casket and delivered to the palace door of the young King Albert.) Later that year, Lever voyaged to the Congo himself, accompanied by his wife and a retinue of staff. Steaming up the mighty waterway, he would have taken in the same primordial-jungle view described by Joseph Conrad in his novella *Heart of Darkness*, based on his own trip to the region two decades earlier: "Going up that river was like travelling back to the earliest beginnings of the world," noted Conrad, "when vegetation rioted on the earth and the big trees were kings." The writer was sick with dysentery and fever for much of his journey, and he witnessed atrocities of the sort described earlier—"saw at a camp[in]g place the dead body of a Backongo," reads one diary entry. "Shot? Horrid smell."—which may explain why, in the book, he describes the Congo River as "an immense snake uncoiled," reaching into "one of the dark places of the earth."

Enjoying the sights from his company's luxury stern-wheeler the *Lusanga* (note the repurposed name), Lever had an altogether rosier take. In Leverville, the oil-palm trees were "all healthy and strong," he wrote in his diary, "and fruiting as well as they can be expected to do in a wild tangle of bush with palms overcrowding each other." He was less sanguine about the workforce. Harvesting oil-palm bunches from natural groves was at least as grueling as—and arguably more dangerous than—gathering latex, and HCB's expat staff had been struggling to find men to do the job.

Despite what his deputies were telling him, Lever couldn't understand why the labor issue was proving such a problem. "The native material is good, willing and anxious to please the white man," he wrote in his diary. "The native African will be a good workman when better understood. I have only praise for him as far as I can judge. Yet the white man here is always speaking of the 'lazy nigger.' He is a child, and a willing child, but he wants training and handling with patience."

Having promised the Belgian government that he would build a school and a hospital in each of his five concessions, Lever made arrangements with Jesuit missionaries to run the former, advising them to train the most promising Congolese as bookkeepers, mechanics, and other tradesmen—so long as they got to them early enough. "It is a well-known fact," he wrote in a letter to one of his directors, "that the brain of the African ceases to be capable of receiving new impressions when he arrives at the adult stage."

Scrambling to fill the growing demand back home for oil and kernels, the Congo's colonial administrators and Lever's agents began pressuring the local chiefs to sign up more of their constituents for work. HCB managers eventually convinced the Belgian government to re-introduce a head tax, and by 1914, all Congolese men were required to pay the *impôt indigène*, with supplemental charges levied on wives. (The latter provision was aimed at the chiefs, who, as polygamists, would be forced to recruit multiple laborers or to send their own slaves to work off their considerable debts.) As in Leopold's day, local agents took to raiding villages, often wielding the dreaded *chicotte.* It wasn't long before they were being met by flying arrows. The recruiters began traveling with military reinforcements and, with the coerced help of handsomely compensated local chiefs, forcing entire villages to relocate closer to Lever's plantations and mills.

Back in England, the businessman held up his African operations as paragons of cross-cultural synergy. "Obviously our only right for existence in the Congo is the development and improvement of the Congo and the native races on sound practical sensible lines," he wrote to a colleague in 1918. "In carrying out this, naturally we do so in a way that will equally benefit the consumer in Great Britain. The two combined make a perfect chain of events which, in my opinion, should be encouraged not only by the Belgian Government but by the British Government."

With the outbreak of war in 1914, Lever's German market collapsed, and the British government began imposing restrictions on all commerce with Holland. "The trade that has taken us nearly thirty years to build up," Lever seethed, "is being wrecked."

In fact, he would profit handsomely from the conflict. In addition to providing munitions makers with glycerine, a by-product of soap manufacturing required for the production of cordite, a gunpowder substitute, he supplied the government with soap and margarine, using cut-price oil and kernels diverted from Germany by the blockade. Between 1914 and 1918, British margarine production rose from 78,000 to 238,000 tons a year, leading Lever to build a new factory devoted expressly to its manufacture. The wartime demand for oil and kernels saw their value skyrocket—palm oil shot from 29 pounds a ton in 1914 to 41 pounds in 1915—eventually prompting the British government to impose ceilings on both.

A continent away, the Congolese reaped little benefit from the

boom. On the contrary, more and more of the laborers being provided by local chiefs were slaves, many of them women and children.
"The Company is finding it hard to recruit a workforce in Niadi *ter-*
ritoire," read a 1921 report by one provincial governor, "on account of
deaths which are supposed to have occurred at Leverville and during
the return journey of certain workers. For this reason, the natives
feel some fear at the thought of enlisting." The distances traveled
and incessant movement between villages and factories were having
the effect not only of spreading disease, he explained, but of breaking
traditional ties of home and family.

In 1923, a Belgian medical officer named Dr. Emile Lejeune conducted a six-day tour of Lever's Lusanga circle and submitted a
report to the governor of the province in which it was located. The
HCB, he wrote, which at the time employed some 6,500 workers, was
importing many of them for three-month stints, but providing them
with neither blankets nor machetes. "Nights are cold in the Kwilu,"
he wrote. "Sometimes a thick mist covers the river bed until late in
the morning and the Europeans for their part have to wrap up warm.
Furthermore, the majority of deaths are due to respiratory ailments.
I regard blankets as an absolute necessity."

He went on to say that the company was by then employing a "very
large number" of teens and children. "I have observed children or
young adolescents at Leverville itself, pushing wagons, and on boats
on the river Kwilu loading timber and fruit. They are not of an age
to do such work." In addition, he wrote, the workers had been denied
the equipment needed for preparing their food. "Anyone familiar
with the abundant portions the blacks ordinarily consume, and the
hygienic conditions under which food is prepared in the villages, will
not be surprised at their dissatisfaction with the diet offered at the
company posts."

While Lever boasted back home of the state-of-the-art housing and schools that he'd built in the Congo, Lejeune described "a brick-built camp, which would be good if there were latrines, kitchens and a rubbish pit, and if it were fenced in, cleared of brushwood and regularly whitewashed. Besides, this camp is only large enough for a very small fraction of the workers currently at the post. . . . One may readily understand how it is that, given such conditions, they refuse after a first stint of three months to re-enlist." At the end of their assignments, he added, the workers had all lost considerable weight. "To sum up, I have found things to be in a deplorable state, and I have been deeply disillusioned by the flagrant practical shortcomings of the HCB's medical service so far as the treatment of its blacks in Leverville and Kwenge is concerned."

(The model garden city of Port Sunlight, meanwhile, would also turn out to be something of a mirage. Fewer than half of Lever's employees and their families actually lived in its houses, and those who did often weren't very pleased about it. "No man of an independent turn of mind can breathe for long the atmosphere of Port Sunlight," wrote the secretary of the Bolton branch of the Engineers' union in a 1919 letter to Lever. "That might be news to your Lordship, but we have tried it. The profit-sharing system not only enslaves and degrades the workers, it tends to make them servile and sycophant, it lowers them to the level of machines tending machines.")

In January of 1924, the interim governor-general of the Belgian Congo forwarded Dr. Lejeune's report to the colonial minister, along with a note: "Bosses in Africa persist in blaming their failure to recruit workers on the indolence of the blacks, when the real cause is to be sought in the fashion in which they treat those in their employ. It is crucial that the Board of the Huileries du Congo Belge at last realizes the real situation in its enterprises in the Kwilu, and intervenes vigorously and rapidly."

The company's intervention would take the form of a defiant doubling-down, with the introduction of something called the tripartite contract, said to have been agreed upon by both the provincial governments of each circle and the local populations. The compacts had the effect of imposing a complete monopoly over the five circles, with local traders barred from selling palm oil or kernels anywhere within a sixty-kilometer radius of each and even prohibited from harvesting their own fruit, being charged with theft if they tried to do so, and disciplined via *chicotte*. Native traditions including dancing, winemaking, and ritual ceremonies were outlawed. A few years later, a Belgian district commissioner would publish an article in the periodical *Congo* in which he described the impact of the tripartite contracts on his own jurisdiction. "At present, the situation closely resembles that prevailing in the concessions of [Leopold's] Congo Free State," he wrote. "Not only the question of land and palms, but also the more or less compulsory work in the forest or in the enterprises of the whites, the *corvées* of various kinds, humiliations such as the ban on hunting, on tapping the palms for wine, on holding ceremonies, all of these innovations and humiliations have disrupted the natives' existence and embittered them," he wrote. "They no longer feel at home 'on the land sold . . . to the company,' and their trust in the justice of the whites has been deeply shaken."

Lever, for his part, was living his best life back in England. In 1920, having previously taken over the last pair of rival British soap companies, he had purchased George Goldie's former RNC, paying more than £8 million to absorb what had by then become an oil-trading colossus. (The news did not go down well at the Nunnery, where Goldie summoned his old colleague the Earl of Scarbrough and offered him twenty thousand pounds cash to go and fight "the soap boiler.") A year later, Lever took a lease on London's Royal Hotel, known as De Keyser's, after the man who'd built it in 1870, and renamed the storied property Lever House, relocating company operations there. Located on Victoria Embankment near Blackfriars Bridge, the building would, by 1932, become too cramped for the ever-expanding company and would be razed to make way for a new Lever House, an edifice the *Daily Standard* described as "the wonder building of modern commerce . . . designed to accommodate 4,000 workers under the most ideal conditions conceivable." At the time, Lever also was furiously constructing various mansions and establishing an art collection to rival any in the country.

In 1924, the mogul made a second and final visit to the Congo. Journeying upriver in a "*Cabine de Luxe*" with wire-gauze netting to shield him from the mosquitoes and a menu featuring *pâté de foie gras*, caviar, "and such like diet to 'rough it' on," as he joked in a letter, he was pleased to see the HCB operating seven mills and no fewer than nineteen steamers and seventy-two river barges. Its African workforce by then numbered some 17,000. "On the foreshore canoes paddled gracefully," he noted in a letter, "no gondola more so, by black-bronze Hercules with dark-skinned matrons and maidens and chubby piccaninnies all happy and smiling."

Smiling, maybe—or at least it might have appeared that way from a distance. Had he spent more time onshore, Lever would have learned that his Congolese workforce had often, like the Brassmen in neighboring Nigeria, felt pushed to the point of breaking. In fact, the nearby country had recently seen another bloody revolt, directed at Goldie's own RNC. In 1929, following rumors that the colony's women would soon be taxed along with its men, demonstrations wracked several Nigerian towns. Police fired on crowds in Jaja's Opobo and elsewhere, killing more than fifty women and injuring dozens more. "Our grievances are that the land is changed," one of the protesters testified afterward. "We are all dying. . . . Since the white man came, our oil does not fetch money. Our kernels do not fetch money. If we take goods or yams to market to sell, Court Messengers who wear a uniform take these things from us." Nigerian history books invariably devote several pages to the country's palm oil–fueled "1929 Women Revolt."

Among the worst of numerous rebellions to break out in the Congo was one that occurred two years later, when a community of Pende tribespeople, inhabiting a place called Kilamba, refused to provide fruit cutters to work in the HCB's Lusanga circle. Receiving word one morning that a Belgian recruiter was on his way to them, the men of the community fled into the forest, prompting company officials to arrest their wives and shut them up in a barn. Over the course of three days, the recruiters and their colonial-functionary pals proceeded to drink a case of booze and to rape the women from the barn repeatedly. By the end of what would turn into a months-long incident, some one thousand Pende had been murdered, and one of the Belgian recruiting agents had been decapitated and dismembered.

In response to the massacre, the Belgians sent a Kinshasa-based lawyer named Eugène Jungers to investigate. "It can be said that virtually all these cutters of fruit were compelled and forced to set out for Leverville," Jungers wrote in his final report, "either by their own 'decorated' chiefs or directly by the civil servants and agents of the territorial service. . . . No 'bushman' knowing something of the tastes and habits of the natives would admit that the latter, when very few things were lacking in their own village, would go and work five or six days' journey away, abandoning for a six-month term their wives and children, in order to live in conditions which are still for all too many of them quite abominable." Out of twenty thousand workers in the service of the HCB, he continued, "scarcely 4,000 live in the magnificent camps set up on the river banks and that, according to a witness, . . . a great number of others, living in wretched huts, are simply kept 'like animals.'"

In 1929, the Niger Company merged with the other large mercantile group in West Africa, the African and Eastern Trade Corporation, to form the United Africa Company, or UAC. (Not to be confused with Goldie's original United African Company.) That same year, Lever Brothers joined with Margarine Union, itself a combine of those two original Dutch firms, to create what *The Economist* described as "one of the biggest industrial amalgamations in European history." Unilever and its subsidiaries did not immediately become poster children for "Being Healthy" and "Being Inclusive." In 1939, for example, a U.K. commission found that the UAC had been acting with other traders to limit the prices paid to West African farmers. During World War II, the Belgian government, then in exile in London, colluded with the Congolese governor-general to impose increasingly harsh demands on the natives for contributing to the war effort. March of 1942 saw them pass an ordinance that doubled, from 60 to 120 days, the length of time that villagers were required to spend in the compulsory cultivation, harvesting, and gathering of oil-palm fruits and wild rubber. They also increased the prison sentence—from seven days to a month—for those failing to comply. For the duration of the colonial period (until 1960), coercion and monopoly remained in place, with village chiefs receiving subsidies for providing laborers, and cutters who refused the work sent off to prison, where the *chicotte* remained in use as late as 1959.

By the time of his own death, in May of 1925 (just three months before Goldie's), Lever had constructed a Gothic church, transformed an old Bolton structure into a museum, and built a handful of mansions, one of them complete with formal gardens, extensive art gallery, and sprawling zoo. He'd also amassed a collection of more than one thousand ethnographic items, from weapons, clothing, masks, and bowls to canoes and native costumes. His patronage of the arts would eventually be compared to that of Henry Clay Frick and John D. Rockefeller, and the author of one book would group the life of William Lever together with those of Andrew Carnegie and Woodrow Wilson. Like Goldie's, Lever's career would be assessed against that of Cecil Rhodes. "In later years the one has often been likened to the other," wrote Lever's son, whose soap-boiler dad had been sure to educate at Eton and Cambridge, "and in many ways their characters and achievements were similar. Courage, organizing ability and breadth of vision were their outstanding qualities, and each did a great work for Africa."

The Africans might beg to differ, of course, but when it came to palm oil, the industry had already begun to move on. A handful of swashbuckling Danes and Frenchmen had set their sights on the increasingly profitable crop and were laying the groundwork for a future in the east that nobody then could possibly have imagined.

4

Playboys of the South China Sea

This jungle will witness other follies.

—Henri Fauconnier, *The Soul of Malaya*

STANDING ON the deck of the *Victoria*, the sea breeze tousling his wavy brown hair, the young Frenchman would have watched as the lights of Marseilles faded into the distance. It was March of 1905, and Henri Fauconnier had just bid goodbye to his mother and four siblings, and to the beloved family home in Charente, some eighty-five miles northeast of Bordeaux. His father, Charles, a charismatic cognac dealer, had passed away four years earlier, bringing to an end an idyllic childhood in the countryside filled with music, literature, and art. Tall and thin, with a narrow face dominated by a hawkish nose, Fauconnier had relocated to England, where he was learning English (and cricket) while teaching music and French to prep-school boys, when he'd happened across an article about Europeans making scads of money cultivating sago, a starch palm, in the Bornean jungle. He'd convinced a wealthy friend from back home to front some money, and the two of them had booked an early-spring passage east.

The ship dropped anchor in Singapore, where the twenty-six-year-old's lasting impression would be of the city's smell: "a thick, warm, yellow odour, hanging in the laden air, that might come from the earth or the throng of tawny humanity on the quay." The guys might have wandered among the docks, crowded with the masts and funnels of ships from across the globe, then splurged for gin slings on the terrace of the Raffles Hotel. In years past, the glamorous property had played host to both Rudyard Kipling and Joseph Conrad, through whose work the bookish Fauconnier had recently been making his way. Three years earlier, Raffles guests had looked on in horror as a tiger, escaped from a circus down Beach Road, had made its way into the Bar & Billiard Room, where it was promptly shot dead.

Word among the expats in town was that it was rubber planters in the Malay States, not sago harvesters in Borneo, who were really cashing in. Three decades earlier, an Englishman named Henry Wickham had wandered deep into the Brazilian jungle and walked back out carrying seventy thousand seeds of the latex-bearing *Hevea brasiliensis*, which he'd loaded onto a ship bound for London. Staff at the Royal Botanic Gardens had raised thousands of the plants and proceeded to send them on to places like Singapore, British Malaya, and the Dutch East Indies. With European financiers backing them, various pioneering expats had established rubber plantations modeled on the lucrative coffee and tobacco operations then dotting the eastern coast of Sumatra and the western Malay Peninsula. Whereas in 1890 some 85 percent of the world's annual supply of rubber was still coming from Brazil (the balance derived from *Landolphia owariensis*, the Congo Basin liana so dear to King Leopold), that was about to change. Harvesting latex from plantations wasn't just incrementally easier than trudging through the rainforest to slice into trees and lianas, it was also, thanks to the low-wage contract laborers streaming in from China, India, and Java, far cheaper and more productive.

In 1905, global demand for rubber sent the price of the commodity to twelve shillings a pound—up from four shillings just three years earlier. Though the sudden rash of planting in Southeast Asia led to a brief oversupply at harvest time, the explosion of the automobile industry around 1910, and the subsequent need for untold numbers of tires, sent prices soaring back up. Before long, the supply of uniform, high-quality rubber flowing out of plantations in the east had doomed the Brazilian industry for good. (Best not to mention the name Henry Wickham on your next visit to Rio.)

Fauconnier secured a six-month apprenticeship on a rubber plantation in Klang, in the coastal lowlands west of Kuala Lumpur, where he learned the basics of cultivation and became proficient in Malay and Tamil, the language of the southern Indians who made up much of the workforce. Within a few months, he was running the place, spending long, hot days directing teams of barefoot "coolies," or hired laborers, as they cleared the forest, established nurseries, and planted young rubber trees in the dirt. (The term *coolie*, now recognized for the racial slur that it is, derives from the Tamil *kuli*, or wages, and appears to have entered the language around the time that indentured laborers replaced slaves across the British Empire.) By the end of the year, Fauconnier was writing to his mother and other relatives in France to request that they wire him money so that he could buy some land of his own. He found a hilly region on the far side of the Selangor River, some forty miles northwest of Kuala Lumpur, and purchased 1,500 acres in a place called Rantau Panjang. For most of 1906, he subsisted in a little thatched hut while building his "House of Palms," the airy, pitched-roof bungalow that he would call home for the next several years. "So much to think about and so much to keep a close eye on," he wrote in a letter at the time. "No furniture, as few clothes as possible, a diet of rice and preserves, the

planter spends the day among his coolies, at once king, judge, doctor, depending only on himself in the solitude of his soul, omnipotent, and alone."

He described the jungle—"with gigantic trees twenty and thirty times the height of a man," fellow expat Madelon Lulofs put it in her 1933 novel *White Money*—and marveled at the amount of time and skill involved in bringing down a single one of its trees. After chopping away the dense undergrowth, a team of laborers would converge with axes to gradually, over the course of several hours, whack, whack, whack at the massive trunk until it finally crashed to the forest floor. In some cases, the root structure would be so vast that the workers would have to build a scaffold in order to access the cylindrical part of the trunk. After felling the trees, they would set fires to clear away the remaining surface vegetation, eventually achieving the sort of dead-brown dirtscape that I'd driven across in Liberia.

While on a port call in Singapore, a Belgian banker and agronomist named Adrien Hallet heard about the adventurous young Frenchman hacking through the jungle north of Kuala Lumpur and arranged to pay him a visit. Hallet, who had grown rich cultivating rubber and oil palm in Leopold's Congo, had recently moved to Sumatra with plans to do the same there. In 1909, he began setting up a range of planting and banking ventures that would eventually grow into his company, La Société Financière des Caoutchoucs, or Socfin (much more about that later), and two years later he made his first pair of oil-palm plantings, one of them at a place called Bangun Bandar, in the island's Deli region. They marked what would become Indonesia's first commercial ventures dedicated to the crop. (The oil palm had been introduced to Southeast Asia some sixty years earlier, when four seedlings were sent to the Botanic Gardens in Bogor, Java, from

Henri Fauconnier in Malaya, circa 1915.

the island of Reunion and from Amsterdam's botanic gardens. They were originally used for ornamentation on tobacco estates.)

A few months after meeting Hallet, Fauconnier traveled to Brussels and, with the financial backing of his new friend, founded his own plantation company. Back in Malaya, he arranged some oil-palm seedlings given to him by Hallet in long decorative avenues leading to the entrance of his House of Palms.

Having quickly made a fortune when the price of rubber tripled—by 1915, the commodity had overtaken tin as Malaya's most valuable export—Fauconnier secured a second plot of land, planting it with coffee and more rubber. He was only three years into the venture when World War I broke out, sending him back to Europe to serve. When coffee and rubber prices tumbled two years later, Fauconnier advised his managers in Malaya to rip out those crops and replace them all with oil palm. That second estate, named Tennamaran, would become Malaysia's first commercial operation devoted to the crop. Fauconnier managed to slip away from his official duties in 1917 to marry Madeleine Meslier, the sister of a fellow planter, at the family home in France. The couple would see little of each other over the next few years, during which time a pregnant Meslier spent several months hunkered down in Saigon. She arrived safely in France with the couple's infant daughter in 1918, despite the ship on which they'd traveled having been torpedoed in the Mediterranean.

By the early 1920s, the ongoing rubber slump had prompted increasing numbers of Southeast Asia's planters to follow the leads of Hallet and Fauconnier. Oil-palm plantations soon covered more than 77,000 acres of Sumatra, and though acreage in Malaya was still limited—just 8,400 acres—it would grow to some 50,000 by the end of the decade. More and more of the oil exported from the region was ending up in the United States, where soap-makers were

working it into their high-end brands, and where margarine manufacturers had begun replacing domestic cottonseed oil with it. The
Americans valued the consistency and mild flavor of the new product, preferring it to the more acidic oil that they'd been sourcing
from West Africa—and prompting British officials there to begin
referring to Southeast Asia's burgeoning industry as "the Eastern
menace."

In the meantime, Hallet was forging ahead on Sumatra, increasing the Socfin acreage and investing in research and infrastructure
projects aimed at ramping up production. He built a bulking installation in Belawan, on the eastern coast of the island, that he designed
to accommodate the giant metal tanks that he'd begun swapping in
for the leaky barrels modeled on those of the African trade. Shipments bound for Europe multiplied, and by 1932, a group of palm oil
producers in Malaya had pooled their resources to build their own
bulking facility, this one in Singapore.

Among the other colorful expats propelling the industry forward
were a pair of Danes named William Lennart Grut and Aage Westenholz. A year after Fauconnier had set sail from Marseille, Grut
stepped off a ship in Siam, where he'd gone to help out his brother-in-
law Westenholz with the latter's new rubber estate. Westenholz had
arrived in the east from Copenhagen two decades earlier, aged just
twenty-six. Already an accomplished civil engineer, he had secured
a position with the Bangkok Tramways Company and proceeded
to rise quickly. With an easy manner and a keen intelligence, he'd
proved a natural leader and had before long been tagged to chair the
board of the Siam Electricity Company. A few years after that, flush
from his city gigs, Westenholz put down some money on a piece of
land in a place called Jenderata, set amid dense jungle some hours

north of Fauconnier's place. (He'd also wired funds to his wayward niece, Karen Blixen, to help underwrite the coffee farm that she was establishing in East Africa.) In 1926, he bought a second parcel of land, at a place called Ulu Bernam, with the express purpose of cultivating oil palm.

Though the new plot was just seventeen miles from Jenderata, the lack of decent roads meant that getting there entailed a seven-hour boat journey up a river infested with twenty-foot crocodiles. Grut and Westenholz spent months slashing through thickets of bamboo and wading across waist-deep rivers, battling not just man-eating reptiles but also wasps and elephants in the course of clearing their land. When a new manager named Axel Lindquist arrived at Jenderata in 1929—by which time the relatives had established a company, United Plantations Berhad, or UP—he learned that a twelve-year-old boy had just been fatally mauled at river's edge. Over the next four years, Lindquist would avenge that killing by dispatching no fewer than 155 crocodiles, some by bullet but most via a technique that he'd dubbed "fishing": he would affix a dead monkey to a large hook and fasten its line to a tree situated along the riverbank. Crocodiles lunging for the primates would find themselves impaled.

It was a random run-in with a Dutch planter on a boat steaming from Bangkok to Singapore that had served as the impetus for the Danes' eventual move into oil palm. Though UP today ranks among the industry's most innovative and successful players, its entry to the sector was far from auspicious. A year after the men had finally put their first oil-palm seedlings in the ground, hordes of rats fled out of the jungle following the onset of the monsoon rains. Westenholz enlisted gangs of Tamil laborers to hunt down the rodents, which had plowed through an astonishing two thousand young palms in two months. What trees did survive reached maturity precisely at the nadir of the inter-war Great Depression.

Rich though many of the expat planters would eventually become, few would have called theirs a life of ease. Arguably as menacing as the heat, the mosquitoes, and the crocodiles were the isolation and the soul-sucking boredom. Retiring to his crude bungalow at the end of a ten-hour day, your average estate man, stripping off sweat-soaked khakis and pith helmet, could basically count on the company of a young houseboy and the comfort of a tepid whisky. In the early years, certainly, when the plantation companies discouraged marriage, European women barely figured in the landscape. Whatever socializing took place amounted to weekend gatherings of the same crowd of sun-reddened bachelors, throwing back warm beers and bitching about their recalcitrant laborers and whatever piece of machinery might have broken down that week. The insularity, the drinking, the racism—all of it would be memorialized in the literature of the time. Graham Greene, reflecting on Malaya under Imperial rule, complained "of British clubs, of pink gins, and of little scandals waiting for a Maugham to record them."

Not just Somerset Maugham, but also the Hungarians Ladislao Székely and his wife, the aforementioned Madelon Lulofs, would base books on the expat planter's life. Fauconnier himself would go on to write a novel, *Malaisie*, or *The Soul of Malaya*, that in 1930 would win the prestigious Prix Goncourt, France's equivalent of the Pulitzer. Its story centers on a character at once enthralled by the exoticism of his new home and tormented by the violence that he's helping to perpetrate—not just on the land but on the people who've always called it home. It's hard to read the book's final "amok" scene and not be reminded of the Delta Brassmen, the Congo's Pende, and the rioting Nigerian women, all of them pushed inexorably to the brink of sanity. "This frenzy that is called *amok* may well be a revenge," Fauconnier's narrator muses, "a self-liberation through revolt; a soul too sensitive to suggestion, humiliated by its own conscious enslavement, at last turns in upon itself, and accumulates so much energy that only the faintest pretext is needed to release it."

The life of a "coolie" was grim indeed. Having fled war and famine in China and India, those desperate thousands enjoyed far less protection from the insects and other innumerable threats posed by jungle existence. Housed in filthy sheds and fed just enough to keep them able to work, they suffered routinely from such ailments as malaria, dengue fever, anemia, and beriberi. Then there were the injuries, whether snakebites, foot lacerations care of the bamboo shards that littered the forest floor, or broken backs resulting from falls out of the palms. As in Lever's Congo, the labor arrangement reflected the thinking of the day. Social Darwinism had posited that the world's races had evolved to form qualitatively different subspecies, each holding a particular place in society. White-skinned Europeans occupied the top rung, with dark-skinned Africans coming in dead last. "From a labour point of view," wrote the British tin-mine mogul Charles Warnford Lock in 1907, "there are practically three races, the Malays (including the Javanese), the Chinese, and the Tamils (who are generally known as Klings). By nature, the Malay is an idler, the Chinaman is a thief, and the Kling is a drunkard, yet each, in his own class of work, is both cheap and efficient, when properly supervised." Like the slaves of the American plantation and the *candomblé*-practicing Afro-Brazilians, the coolies looked to both music and spirituality as a source of comfort and means of resistance. "I hoe all day and cannot sleep at night," went one of their work songs. "Today my whole body aches, Damnation to you *arkatis* [recruiters]./I have toiled day and night/from the moment I entered your house./The skin of my body has dried/And happiness has become but a dream."

By 1934, the Dutch East Indies (today's Indonesia) had surpassed Nigeria in annual production of palm oil, an event that signaled a seismic shift in the industry. A few years earlier, when officials from the colonial office in the latter had sent a pair of agricultural experts east to figure out how they'd been so disastrously outrun, they couldn't have been pleased with the length of the report that came back. For starters, they were told, the oil palms in Sumatra and Malaya had been planted in neat rows, with wide paths left between them to facilitate accessing the plants and maintaining the surrounding terrain. Spaced a good distance apart, the trees grew more slowly and their bunches matured closer to the ground, making it easier and far less dangerous to harvest their fruit. Plantation owners in the east had begun to install narrow-gauge railways, whose cars carried the bunches directly from field to mill, helping to keep acidity low. (Later, railway tank wagons outfitted with steam-pipe coils would ensure that the oil stayed liquid for pumping directly into steamers waiting in the ports.) In short, the various upgrades and quality-control systems that had been put in place in Southeast Asia suggested that its palm oil industry—abetted, naturally, by filched genetic material and a system of virtual slave labor—would see the Nigerian trade go the way of Brazil's rubber business. Which is to say, kaput.

But that would take some time. The outbreak of World War II had the effect of slamming the brakes on the region's nascent industry, which by that point had grown to some 235,000 acres of plantings in Sumatra and 57,000 in Malaya. When, in 1941, Japanese troops descended on both, the European planters fled—a few of them making near-death escapes behind the controls of their own planes—leaving their Chinese and Tamil workers behind. Plantations were occupied and managers imprisoned, with several hundred workers being marched off to labor on the "Death Railway" in Siam. Only half ever made it back. (The French novelist Pierre Boulle, who worked as a planter for Hallet's Socfin both before and after the war, wrote about the railway in his novel *The Bridge on the River Kwai,* later the basis for the award-winning film.) Returning to their estates in 1945, the Europeans found that their crops and equipment had been destroyed, though the Japanese had spared some of the oil palms, apparently prizing the fruits for their high vitamin content.

Among the places forever marked by the war would be the port town of Sandakan. The capital of British North Borneo (now Sabah) until 1947, the city has in recent years earned a reputation among adventurers as a jumping-off point for jungle treks to see orangutans and pygmy elephants. But the cheery tourist brochures tell only part of the story. The airstrip on which I touched down during my own visit, I learned, had replaced another one built half a century earlier by Australian and British prisoners of war brought there in 1942. The men, who originally numbered some 3,500, had been lodged at a camp a mile and a half away. At the end of 1944, upon orders from their superiors to "Annihilate them all," Japanese soldiers shot, hung, or poisoned one thousand of the prisoners in Sandakan. Starting in January of 1945, by which time Allied raids had rendered the airstrip unusable, the Japanese marched the remaining 2,434 captives to Ranau, 165 miles inland. Emaciated, clothed in rags, and suffering variously from dysentery, malaria, and beriberi, the prisoners trudged through swamps and up mountains, hacking their way through thick jungle and following tracks that led straight across rivers. Those who dared take a break were shot, bayoneted, or clubbed to death. (There were later reports of crucifixions, castrations, and cannibalism along the route.) By the end of 1945, every one of the Sandakan prisoners was dead, save for six Australians who had managed to escape. The photographs on view at the Sandakan Memorial Park, located on the grounds of the former prisoners' camp, could have been taken at Auschwitz.

(A continent away, the war did beget at least one happy legacy: at the height of the fighting, a baker in Piedmont, Italy, had begun working his region's plentiful hazelnuts into chocolate confections as a way to stretch his wartime cocoa rations. Pietro Ferrero originally sold his *Giandujot* in solid blocks but introduced a creamy version in 1951. A decade later, his son took over the business, fiddling with the recipe for what had become known as Supercrema and renaming it Nutella. In the years since, the family-run company has become increasingly reliant on palm oil, which today makes up fully one-fifth of Ferrero's universally adored spread.)

Prior to the outbreak of the fighting, agronomist Hallet had been tinkering with oil-palm breeds in hopes of developing a crop that was both hardy and capable of producing copious oil. Based on those original plants from the botanic garden, his palms were mostly of a variety known as *dura*, which yielded fruits with a thick shell and a relatively pulpy fruit. Having evolved in response to Sumatra's particular soil and climate, they had come to be known as Deli *dura*, after the location of Hallet's estate, and had eventually been adopted across the region. But Hallet was convinced that he could do better. Back in 1902, German botanists working in Cameroon had identified an oil palm known as *tenera* that produced fruit with a comparatively small kernel and an even higher ratio of oily flesh. Found rarely in the wild and unreliable when it came to breeding, the variety had been mostly forgotten until the 1930s, when researchers in neighboring Congo had determined that *tenera* was actually a cross between the *dura* and a rare, shell-less variety known as *pisifera*. With this knowledge in hand, Hallet had begun crossing *dura* with *pisifera* in pursuit of a *tenera* suited to Sumatra. His trial plantings were in the ground at Socfin's Bangun Bandar estate when the war interrupted everything.

In the years after the conflict, synthetic rubber continued to erode the market for the natural stuff, leading European plantation owners to replace more of their remaining rubber with oil palm—increasingly, with Hallet's *tenera*. But restoring the industry to its pre-war vigor proved a complicated enterprise. Part of the fallout of the fighting had been a galvanized anti-colonial sentiment, particularly among locals, who had already seen their way of life upended by the multiplying plantations. During the years of the conflict, Chinese laborers had taken over abandoned estates and opened up new land. They lacked legal rights, however, and the returning Europeans, dismissing them as "squatters," were now telling them they had to go. By 1948, a guerilla insurgency born of these and other grievances had broken out in the rural areas of Malaya. (A similar storm was brewing across the Malacca Strait on Sumatra, a situation we'll explore in Chapter 5.) The Malayan Emergency, as the conflict came to be known—at least among the Brits; the Malayan National Liberation Army called it "the Anti-British National Liberation War"—would simmer on for a dozen years, the rubber and oil-palm plantations serving as its central staging grounds. City-dwellers at the time were said to have been oddly insulated from the conflict. Such urbanites, writes Christopher Hale in his 2013 book *Massacre in Malaya, Exposing Britain's My Lai*, would "[stare] unbelievingly when grim-faced planters or plantation owners strode into hotels in Kuala Lumpur like characters out of a

Wild West film with revolvers tucked into their waists or grasping formidable looking rifles."

In fact, the Emergency would claim the lives of several such planters and plantation managers, and, in 1951, of the British High Commissioner himself. Fifty-three-year-old Henry Gurney and his wife were seated in the backseat of their black Rolls-Royce when their car was ambushed en route to a golf club located sixty miles north of Kuala Lumpur. Gurney was gunned down when he tried to run from the car. An earlier massacre, on a rubber plantation at a place called Batang Kali, had involved a village torched to the ground and the mutilated bodies of twenty-four civilians.

With the Emergency raging in the background, Malaya's colonial officers and plantation companies put their best efforts into nudging the industry forward. In 1955, the British company Harrisons & Crosfield—recall the second name from Lever's patent-infringement lawsuit—created a research station south of Kuala Lumpur to focus on breeding and pest control. Hallet's Socfin, having since absorbed Fauconnier's company, established a similar facility, as did the Danes over at UP. Back in the Congo, Lever's old HCB, by that time a subsidiary of Unilever, had developed a screw-based mill designed specifically for the extra-pulpy *tenera* fruits that had been quickly adopted across Sumatra and Malaya. With its expanding soap and food operations requiring ever-greater volumes of palm and palm-kernel oil (and with the Congo beginning to descend into political turmoil), Unilever, too, began looking east, establishing some eleven thousand acres of oil palm in Kluang, north of Singapore, and eventually adding land in Sabah, on the northern tip of Borneo.

"Colors, moisture, heat, enough blue in the air." Such were the reasons offered by Inez Victor, the protagonist of Joan Didion's *Democracy*, when asked by the novel's narrator why she had opted to stay on in Kuala Lumpur. Back in 1975, the year the novel takes place, the Malaysian capital was still something of an obscure backwater. ("We were sitting in a swamp forest," writes Didion, "on the edge of Asia in a city that had barely existed a century before and existed now only as the flotsam of some territorial imperative.") In the four-plus decades since, all that has changed. Today a visitor to Kuala Lumpur can check in at a Ritz-Carlton, a St. Regis, or a Four Seasons, and she can stride out of the gleaming-white shopping mall toting bags from Bottega Veneta and Burberry, Tiffany and Tory Burch. The city's skyline, including the twin phalli that are César Pelli's Petronas Towers, until 2004 the tallest buildings in the world, announces itself with all the subtlety of a Trump-branded hotel. Cranes hover at right angles all across town, where bright-green netting drapes over the skeletons of countless skyscrapers-in-the-works.

But descending over Kuala Lumpur today, the view isn't so much of colors-plural as it is of a singular, distinctive shade of green. Oil-palm plants spread out below in all directions, millions upon millions of tiny emerald asterisks lining up in neat matrices or cascading down endless terraced hillsides in loose concentric swirls. So extreme has Malaysia's mania for oil palm become over the decades that the thick-trunked plants now sprout within mere yards of the international airport's runways. Driving away from the sleek building, all space-aged curves and cantilevers, you get a ground-level take on the very same vista: oil palms, planted right up to either side of the eight-lane highway.

After independence, in 1957, the Malaysian government faced a challenge in redistributing the country's wealth among its seven-million-plus citizens. With most of its rural areas still desperately poor, it established something called the Federal Land

Development Authority, or FELDA, the main aim of which was to provide "land for the landless, jobs for the jobless." Its anchor program entailed the clearing of vast areas of rainforest to make way for the resettlement of hundreds of thousands of poor Malay families. In addition to small plots of land, the newcomers were given houses and rubber and oil-palm seedlings. When, a few years after its inception, prices of the former fell again, FELDA moved more deeply into oil palm. In the decades since, the agency has encouraged partnerships with foreign plantation companies and has developed hundreds of thousands of acres of its own oil-palm estates, both across Peninsular Malaysia and on neighboring Borneo. (The world's third-largest island, Borneo is divided among Malaysia, which occupies the top third or so, and Indonesia, with a little wedge cut out of the north for the sultanate of Brunei.) Today, FELDA ranks among the largest oil-palm landholders in the world, operating both commercial private estates and smallholder plantations in addition to a diversified line of

Oil-palm deforestation in Malaysia.

businesses that include extracting, refining, trading, and marketing consumer goods, plus various unrelated ventures.

By 1966, Malaysia had become the world's leading exporter of palm oil, overtaking neighboring Indonesia. Among the places its industry found to park all of that product were the convenience-food and catering businesses that had emerged during wartime Europe to feed the continent's soldiers and factory workers. With refiners having since figured out how to transform the brightly hued and strong-tasting substance into something odorless, flavorless, and all but invisible, these new processed-food manufacturers had come to rely on it for replacing the more expensive butter and lard, while providing texture and shelf life to their growing lines of cookies, crackers, and other baked goods. Its high smoke point rendered palm oil ideal for frying up mass quantities of potato chips, or "crisps."

The oil also was making inroads into home kitchens. In the late 1950s, the international medical community had begun to warn that the high saturated-fat content of butter appeared to increase the risk of heart disease, prompting consumers across the globe to seek out vegetable oil–based margarines. By 1969, the United Nations was reporting that people in eighty-five countries across Europe, the Americas, Asia, and Oceania were now using industrial vegetable oils, including those made from oil palm, to fry their foods rather than boiling them, as they previously would have done.

In 1974, the Danes at United Plantations opened Malaysia's first industrial palm oil refinery. (UP management also gets credit for having introduced, some years earlier, the harvesting pole, which quickly became the norm on plantations worldwide.) No longer would all of those tankers shipping out from Malaysian ports be carrying crude oil destined for processing on foreign shores, and no longer would the considerable profits from the costlier product bolster only bank accounts abroad. Other plantation companies followed with refineries of their own, just as the industry's technicians found novel ways to massage the oil into variants with textures and chemical profiles to suit seemingly any need. The spike in crude-oil prices during the 1970s also proved a boon to the industry, with new, palm oil–based oleo-chemicals filling the void created when petrochemicals manufacturers pulled back supplies. European countries increased their imports of palm oil, and Malaysia responded by putting still more plants in the ground, including through further expansion on Borneo. Between 1964 and 1984, during which time Malaysia began a partial nationalization of its plantations sector, the country's oil palm production grew more than twenty-five-fold. In 1979, the government established something called the Palm Oil Research Institute of Malaysia, or PORIM, tasked with promoting the industry, and twenty years later, PORIM merged with another state body, the Palm Oil Registration and Licensing Authority, to become the Malaysian Palm Oil Board, or MPOB. Remember the name: we'll track the organization's exploits in the chapters to come.

The early 1980s brought yet another important innovation to the Malaysian industry, this one in the form of a tiny bug. Up until then, pollination across Southeast Asian oil-palm operations had been done manually, with armies of workers wielding specially designed bellows to carefully blow pollen onto female flowers at the exact opportune moment. But a Scot named Leslie Davidson, a veteran Unilever man, had long suspected there might be a better way. Having moved to Malaysia as a twenty-year-old, Davidson had by 1951 been cultivating oil palm for decades, including during a stint at the company's operations in Cameroon. While on the ground in West Africa, he had noticed that insects regularly swarmed around the oil-palm plants. And so in 1974, by which time he'd ascended to vice-chair of Unilever International Plantations Group, Davidson sent three entomologists back to Cameroon to investigate. Sure enough, the scientists determined that weevils of the species *Elaeidobius kamerunicus* served as the plant's main pollinators. *Elaeidobius* didn't exist in Malaysia, but after extended negotiations with the government there, Unilever secured permission to import mass quantities of the African bugs. On February 21, 1981, staff at its Johor estate giddily released some two thousand of the insects into the air. Within a year, the country's annual oil-palm yields had shot up by 400,000 metric tons.

With production volumes headed ever upward, the industry began to face challenges unloading all of that supply. Companies teamed with the Malaysian government to establish new markets in Asia, Eastern Europe, and Africa, and they continued to find new ways of adding value to their product. The larger ones integrated vertically, not just refining the oil that they'd produced but also crushing the kernels and manufacturing cooking oils, food products, specialty fats, oleo-chemicals, and, eventually, biofuels.

I pulled on a hard hat, zipped up a fluorescent-yellow vest, and slid into a pair of chunky black shoes, then followed a technician through the lobby of the Sapi Oil Palm Mill and into the lot out front. Located in a place called Beluran, a two-hour drive west of Sandakan, the mill processes some thirty thousand tons of palm oil every year. A dozen dump trucks piled high with spiky bunches sat lined up outside, where a smell that brought to mind burnt molasses filled the morning air. Still more bunches lay mounded on the ground, their fruits gleaming tangerine and crimson in the sun. It was mid-October, the height of a harvest season here, so the action at Sapi, one of more than 450 mills now dotting the country—oil-palm plantations cover some 14 million acres of Malaysian land—extended around the clock.

Ready for processing at the Sapi Oil Palm Mill, in Malaysian Borneo.

One by one, the trucks drove onto a pair of weighbridges, where agents in air-conditioned booths punched numbers into computers. The drivers then dumped their loads into a series of metal rail cars that progressed along tracks to a pressure cooker–like device that both sterilized the fruits (halting the development of free fatty acids) and loosened them from their husks. Back inside the building, where the roaring, whooshing, clanking machines reduced my technician-guide to shouting, the individual fruits got stripped from their bunches and threshed around in a drum, emerging with the look of charred dates. Operating like a slow-speed blender, a digester then loosened the flesh from the nuts and heated the whole mass, readying it for the press, which expelled a thick, blackish sludge of oily fiber and nuts. The sludge was then fed into a centrifuge, which separated it into a deep-brown "decanter cake" and an oil that, once clarified, got funneled into steel tanks, ready for transport to the refinery. The kernel-containing nuts, meanwhile, sporting a stringy beige fringe that gave them the look of tiny coconuts, would get trucked off to different facilities to be crushed and their kernels extracted.

A few days later, having successfully navigated the roundabouts and honking midday traffic of the industrial dump that is modern Sandakan, I settled in at a conference table at Sandakan Edible Oils, a refinery owned by Wilmar International, the largest palm oil trader in the world. (I'd confirmed access to the high-security facility some weeks earlier, conversing by email with the media representative of the Singapore-based company.) Three middle-aged guys in logoed, short-sleeved shirts walked me through the refining process, illustrating the stages and resulting products with a series of jars con-

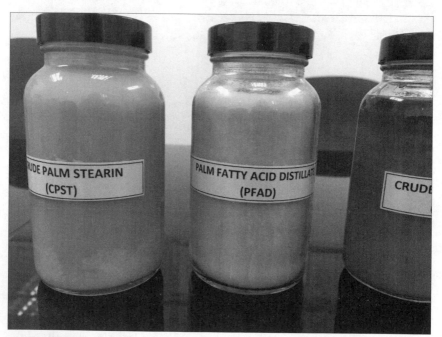

Palm oil in various stages of refinement at Sandakan Edible Oils, in Malaysian Borneo.

taining palm and palm kernel oils of various colors and densities. Treatment with phosphoric acid, they explained, at temperatures ranging from 90 to 110 degrees Celsius, served to de-gum the thick crude oil, after which it got bleached, cooled, and filtered. Finally, it would be "steam stripped," at temperatures of up to 270 degrees, to remove free fatty acids and volatile compounds, a process known as neutralizing and deodorizing. The resulting "RBD," or "refined, bleached, and deodorized" oil, makes up much of what gets sold on commodities markets worldwide.

Some of the bigger refineries further "fractionate" the oil into solid and liquid forms, known as stearin and olein, respectively. Much of the former gets shipped off to oleo-chemical plants, where it's further manipulated and broken down into various fatty acids, fatty alcohols, esters, and glycerines, which in turn are sold to detergent and cosmetics manufacturers, and to the chemical industry. Almost all of the olein ends up with the food industry, to be sold as cooking oil or used in processed foods. Palm kernel oil, having also undergone bleaching and deodorizing, eventually moves on to oleo-chemical companies, which fractionate it for use in cosmetics and personal-care products. (The palm kernel meal that remains, known as expeller, mostly ends up with the livestock industry, which prizes it as a cheap source of protein.) At this point, the palm oil has very little in common with the traditional foodstuff native to West Africa and Bahia, Brazil. It's gone through so much processing, in fact, that it can turn up in products listed under any of some two hundred names.

In 2019, thanks to decades of development and aggressive marketing, Malaysia's palm oil industry produced a record 20.5 million metric tons of oil, worth some $9 billion. But in the last two decades, the country has lost no less than 20 million acres of tree cover.

Henri Fauconnier wouldn't recognize the place. Having left Mala-ya in the mid-1920s, the Frenchman settled with his family first in Tunisia and then back at his childhood home in Charente, where he enjoyed life as a celebrated literary figure, exchanging letters with the likes of Jean Cocteau, Colette, and André Gide. On his deathbed in 1973, the longtime planter made a final request to the relatives who had gathered around: that they put on a recording he'd made many, many years earlier, of the mellifluous wall of sound that was the Malayan jungle at night, circa 1925.

PART II

SOMETHING'S ROTTEN

5

Silent Summers

There was a strange stillness.
The birds, for example—where had they gone?
Many people spoke of them, puzzled and disturbed.

—Rachel Carson, *Silent Spring*

THE ONE with the gun arrived with a cocky flourish, sauntering through the doorway in a white muscle tee and blue jeans torn at the knees. He settled in on the plastic flooring, lit up a thick clove cigarette, and began to talk animatedly about the twenty-three critically endangered birds that he'd shot from the Indonesian sky in the previous five months. All cheekbones and shiny black hair, the handsome thirty-seven-year-old passed around his 4.5-millimeter weapon so we visitors could admire its sleek caramel stock and gleaming brass barrel. He was happy to be photographed and videotaped mimicking the high-pitched-honk-transitioning-to-maniacal-laughter of his feathered victims. *"Koo! Koo! Koo! Koo! Koo! Koo! Koo! Koo! Koo-koo-koo-koo-koo-kah-kah-kah-kah-kah-kah-kah-kah-kah-kah kah!!!"*

As recently as a few years ago, neither this guy nor either of his pals could have picked out a helmeted hornbill from a children's picture book. Now the three went on about how the bird they call *rangkong* travels in pairs and favors the high branches of a particular fig tree across the river. The birds come out in the morning, they said— around seven or eight—and again in late afternoon. When a male is killed, his mate will appear "a bit lost," flying around in search of him and calling out to her friends.

By the time I arrived in northern Sumatra, poachers' guns were just the latest threat facing the region's storied birdlife. Since the days of Adrien Hallet, tens of thousands of square miles of rainforest have fallen here to make way for oil-palm cultivation. As the forests have disappeared, hornbills, as well as orangutans and other creatures, have found themselves squeezed into ever-tinier patches of suitable habitat. Today more than 75 percent of Sumatra's 102 lowland forest–dependent bird species are considered globally threatened. At the same time, plantations and the new roads that go with them render what's left of the forests that much more accessible to poachers like these guys, who kill the helmeted hornbills for their casques, solid-keratin enlargements on the upper part of their bills. Long prized by the Chinese for sculpting into snuff bottles and jewelry and grinding into traditional medicines, the casques have taken on new status in recent years, thanks in part to the growing difficulty of procuring elephant tusks.

The wildlife-trade watchdog TRAFFIC recently reported that some 2,878 helmeted hornbill casques and skulls were confiscated in the region between 2010 and 2017. Having decimated populations in Kalimantan (the Indonesian part of Borneo) and in southern Sumatra, the poachers had now shifted their efforts north. In 2016, authorities in Aceh province, here at the northern tip of Sumatra, confiscated twelve casques, two rifles, a digital scale, and disposable cellphones from a pair of Acehnese men who confessed to having sold

Poacher in Aceh province, on the Indonesian island of Sumatra.

at least 124 beaks to Chinese middlemen in the previous six months. It was in 2018 that the Switzerland-based International Union for Conservation of Nature, or IUCN, raised the alert level on the helmeted hornbill to "critically endangered."

Yokyok "Yoki" Hadiprakarsa, a conservation biologist who directs the Java-based Indonesia Hornbill Conservation Society, explained to me that Indonesians who for centuries had lived off the land, sourcing their food, building materials, medicines, firewood, and water from the forests, were now finding themselves having to pay for such necessities. That, in turn, had led to the recent epidemic of hornbill poaching. "People fight on a day-to-day basis to fulfill their daily needs," he said, "so they look for quick opportunities. For those living near the forest, hunting for wildlife is the obvious option."

On August 17, 1945, two days after Japan surrendered to the Allies, Indonesia declared independence from the Netherlands. It was time:

Jakarta, then known as Batavia, had been founded by the Dutch more than three hundred years earlier. A young architect named Sukarno (Indonesians often go by a single name), a prominent member of the country's nationalist movement, assumed the presidency and proceeded to lead the country under a policy he called "guided democracy"—informed simultaneously by nationalism, religion, and communism. (Though the Dutch had outlawed the Communist Party, or PKI, in 1927, it had continued to flourish underground in the years leading up to the war.) Sukarno nationalized foreign companies and expropriated Dutch property, including oil-palm plantations, and, over the years, largely at the behest of the Communist Party and the plantation-workers union, he'd sought to enact land reform and to improve wages and living conditions for the country's plantation workers. The initiatives were not uniformly applauded. Landowners, plantation managers, and middle-class Indonesians in particular fought to hold on to whatever power and property they could. Meanwhile, communist sympathy was surging: by the early 1960s, the PKI had become the largest party in the country, claiming a membership of some 20 million.

In October of 1965, the country's long-simmering social tensions came to a devastating head when an alleged coup attempt served as impetus for the Indonesian army and its related paramilitary groups to slaughter between 500,000 and one million members of the PKI and its purported associates. Over the course of five months, grisly massacres played out across the archipelago, with death squads moving from village to village and murdering every supposed communist, trade unionist, and peasant in their paths. Also targeted were Chinese immigrants, amid allegations that Beijing had backed the alleged coup. Among the main settings of the atrocity, which the CIA called "one of the worst mass murders of the 20th century," were the oil-palm and rubber plantations that stretched across north-ern Sumatra. (Documentary evidence of the CIA's own complicity in the event has since been widely acknowledged.) The Sungai Ular, or Snake River, located just ten miles from Hallet's original estate at Bangun Bandar, figured prominently in the genocide, clogged as it was nightly with bodies washing out to sea. (The estate is still owned by Socfin today.) An executive from a Goodyear rubber plantation located nearby would later describe what had become of the hun-dreds of plantation workers who'd been rounded up and detained over those many months: "Every Saturday night, a couple of trucks arrived and took away a hundred or so [people] to the nearby bridge by a fast-flowing river close to the plantation headquarters. They were killed with jungle knives on the bridge and their bodies were thrown in the river."

By 1967, when a general named Suharto was named acting presi-dent, the Indonesian population wasn't just traumatized—it was also deathly poor, with 60 percent of its citizens surviving on less than a dollar a day. Over the next two decades, the World Bank would advise Suharto to expand the palm oil sector as a way to engage its rural poor, just as the Malaysian government had done a decade ear-lier. In the late 1970s, Suharto introduced something called Trans-migration, another World Bank–backed scheme, which entailed the forced resettlement of millions of Indonesians from the archipelago's crowded islands to less-populated ones like Sumatra and Borneo, where they were encouraged to cultivate the crop. In subsequent years, the government would launch campaigns aimed at promoting domestic consumption of palm oil. Whereas in 1965 it accounted for just 2 percent of Indonesians' cooking-oil usage, by 2010 that figure had soared to 94 percent. The World Bank also underwrote Suharto's forest policy, under which more than half of the country's

rainforests were logged and converted to plantations. The folks on
the receiving end of those lucrative logging and oil-palm concessions? The president's family, friends, and fellow military officers. In
2004, Transparency International ranked Suharto the most corrupt
leader of all time.

As with Malaysia's FELDA schemes, Transmigration entailed
the clearing of massive amounts of land and the displacement of the
peoples who had lived and hunted there for generations. In the last
few decades, Sumatra's semi-nomadic Orang Rimba and Batin Sembilan tribes have lost tens of thousands of acres of forest to the palm
oil industry. Though Indonesian law recognizes the customary land
rights of its indigenous peoples, development projects deemed to be in
the national interest—such as logging concessions and plantations—
have long superseded them. Then as now, the sort of land grabbing

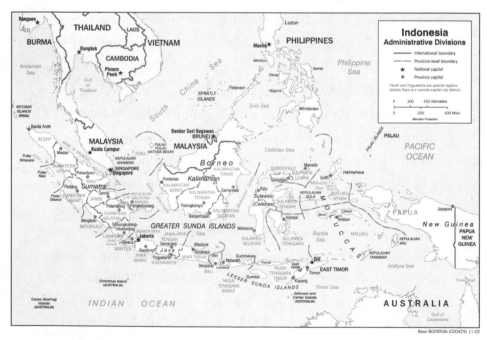

Present-day Indonesia, with provinces delineated.

that I'd gone to investigate in Liberia is here largely perpetrated by the state itself (though often with the encouragement of outside governments and financial interests). During 2016 and 2017, according to the Ombudsman Republik Indonesia, which investigates maladministration in the country, oil-palm plantations were the source of more land conflicts than any other sector. In 2018, Ombudsman recorded more than one thousand land complaints, many of them from indigenous communities, against Indonesia's palm oil companies.

As the hornbill situation illustrates, the loss of land and livelihoods by indigenous communities serves to further decimate biodiversity, a crisis with ramifications well beyond Sumatra. The 2019 U.N. report I mentioned in the prologue found that as many as one million species of plants and animals are today threatened with extinction. The situation is particularly bleak for tropical forests, which, recall, house more than half of the world's biodiversity. The demise of individual species can lead to the collapse of entire ecosystems, impacting not just local communities but ultimately destabilizing economies and governments and triggering famine and refugee crises.

A violent thirty-year separatist insurgency had long spared Aceh the grim environmental fate of the rest of Sumatra, but the signing of a 2005 peace accord put an end to that. Since then, the palm oil industry has set its sights on something called the Leuser Ecosystem, a 5.6-million-acre expanse of lowland and mountainous rainforest that spreads across the bottom half of the province. Home to 382 bird, 105 mammal, and 95 reptile and amphibian species, the butterfly-shaped Leuser is a UNESCO-designated World Heritage site and ranks among the most biologically diverse places on Earth. (The poachers live at its heart, in a hamlet called Tamiang.) The Leuser, one-third of which forms Gunung Leuser National Park, is the last place on the planet where there is terrain of sufficient size and quality to support viable populations of Sumatran tigers, elephants, and rhinos, and of orangutans, clouded leopards, and sun bears. In addition to its helmeted, rhinoceros, and other hornbill populations, the Ecosystem is alive with the calls of the tan-breasted partridge, the salvadori's pheasant, various laughing thrush, and the critically endangered Rück's blue-flycatcher.

Considered a National Strategic Area for what the government terms its "environmental-protection function"—its forests provide a steady, clean water supply to more than four million Acehnese—the Leuser is technically safeguarded under Indonesian law. Still, the past decade and a half have seen roughly five thousand acres of its park converted to oil-palm plantations. Today, only 4.5 million acres of the Ecosystem remain forested. Here as elsewhere in Indonesia, palm oil companies have secured permits through backroom deals with local officials or have simply paid others to clear the land illegally. (We'll get into the details in Chapter 8.) Political connections and a lack of oversight mean they largely do it with impunity.

It was at the end of 2014, explained the oldest of the three poachers, the one who'd invited us into his house and served us syrupy coffee, that he and the others had begun to notice strangers in their midst, men from Jambi province, south of this remote village, and Chinese fellows who'd swoop in for a day or two before disappearing again. Eventually it became clear that the outsiders had come in search of helmeted hornbills, one of ten hornbill species that make the island their home. (Among about sixty hornbill species worldwide, roughly half are native to South Asia. The other hornbills, none of which is seriously threatened, are endemic to sub-Saharan Africa.) At $6,000 a kilogram, the birds' casques, commonly referred to as "golden ivory," "red ivory," or "golden jade," were selling for five times as much as elephant ivory. Hong Kong shops displayed intricately carved trinkets made from the body part with price tags in the tens of thousands of dollars.

In addition to diminishing the habitat of Indonesia's hornbills, the incursions have impacted the particular living requirements of the birds. Known as the "farmers of the forest" for the critical role they play in dispersing seeds, hornbills need dense habitat and a steady supply of fruit. Their unique nesting habits depend on the sort of old-growth trees that tend to fall first to developers. A female ready to lay her eggs retreats to a natural cavity inside a wide trunk. She and her partner seal the entrance with a paste of fruit, mud, and feces, leaving a small slit through which he will feed her (and, eventually, her chick) for up to five months. Killing a male hornbill, then, as poachers are wont to do given the gender's larger size, often means the demise of female and chick as well.

I'd landed in that modest brick home thanks to a local environmentalist named Rudi Putra. A forty-three-year-old with a faint goatee and a degree in conservation biology, Putra had developed an early love for his native island's iconic rhinoceros and had several years earlier decided that he'd devote his life to protecting it and the region's other wildlife. It was a calling that often involved run-ins with poachers like these.

Sumatra's once-sheltered rhinos, like the hornbills, increasingly were being squeezed out of the Leuser's forests. Along with its elephants and orangutans, they had begun encroaching on local communities as a result. Farmers and plantation workers, annoyed by the beasts' habit of knocking down homes and trampling through crops, had been responding by setting out traps or potassium cyanide–laced pineapples, or by shooting the animals with pellet guns. Once wide-spread across Southeast Asia, Sumatran rhinos were now down to an unimaginable eighty individuals. Snakes, too, were getting caught up in the human-wildlife standoff. In 2017, a young father had left his home to harvest oil-palm fruits on the neighboring island of Sulawesi and was discovered hours later by a group of villagers when they sliced open a bloated twenty-three-foot reticulated python crashed out on the plantation's edge. The snakes don't typically prey on humans, but with so few mammals remaining in their midst, they'd begun to get desperate.

Putra traced his own awakening to the importance of tropical rainforests—and, in particular, of an intact Leuser Ecosystem—to 2001. That year, while working as a government-funded researcher there, he had witnessed a violent flood ravage his and other down-stream communities, an event widely attributed to the region's recent deforestation. When support for Putra's research job dried up, he took to arranging meetings with community members, cops, local officials, and civil-society groups in an effort to thwart the palm oil industry, which by 2000 had supplanted logging as the number-one threat to the Leuser. He began leading teams of volunteers into the forest to confront poachers and dismantle their snares, and he eventually founded the Leuser Conservation Forum, which today employs more than seventy rangers to keep watch over the region, shielding it from illegal palm plantations and poachers alike. In 2014, Putra received the Goldman Environmental Prize, a $175,000 honor given annually to a handful of grassroots activists working across the globe.

Slim bordering on concave, and with an ascetic fashion sense that runs to rubber flip-flops and worn T-shirts, the soft-spoken father of two is an odd mix of high-functioning executive, self-contained prophet, and guileless child. Wielding a flat Samsung phone in one hand and a skinny Nokia in the other, he expertly juggled calls from multiple continents but dissolved into giggles when recalling his initial face-to-face encounter with a rhino. (They ran in opposite directions.) At a meeting in an open-air bungalow set high on stilts, Putra sat cross-legged before twenty-six employees dressed in "Wildlife Protection Team" garb and spoke with a quiet authority about the group's often-dangerous work. In a country where a lit cigarette is an arguably more de rigueur meeting accessory than either a notebook or a pen, he quietly commanded one ranger to kindly snuff his out. "Remember that the people who go into the forest to poach are our friends and family," he counseled the members, who ranged in age from twenty-five to seventy. "We shouldn't hate them. We should be gentle with them and explain why they shouldn't poach."

By 2009, Putra, who still accompanies his rangers on fifteen-day patrols each month, had begun taking chainsaws to northern Sumatra's illegal oil-palm plantings. (Provincial officials issue preliminary "scoping" permits under which companies are expected to secure the consent of local communities and prepare environmental assessments as steps toward eventual permission for development, but many forge ahead without following through on either.) We spent one cloudy afternoon traipsing over hills and weaving among gigantic palm fronds to reach a stand of trees in a 2,600-acre plot on the eastern fringe of the Leuser. Trailed by a handful of curious kids and accompanied by eleven local guys toting banana, durian, and other seedlings—they sow native crops on the sites where they've downed the oil palms—we looked out across a terraced landscape of muddy green, the only variation in hue the pale moons of already felled trunks. One of the men revved a chainsaw and then drove the machine's five-foot blade through the base of a fat trunk, sending the behemoth crashing to the ground. Though we encountered no resistance on that particular day, Putra, who at that point had dismantled twenty-six illegal plantations—some 7,500 acres of oil palm—said that confrontations are a part of the job. The support of local cops helps, he said, but he and his colleagues routinely face off against locals and company officials, one of whom sued him for damaging his property. (In a rare legal win, the plantation in question had been deemed illegal and its owner ultimately forced from the land.)

Back outside the plantation, I stopped to chat with a local named Ngatimen. In the late 1990s, he told me, he and some fellow villagers had planted oil palm in a previously logged section of forest. "We didn't do a cost-benefit analysis," he said. "We thought we'd sell the fruit easily." When, in 2012, global palm oil prices tanked, the villagers found themselves struggling to feed their kids. They've since replaced the palms with lemon, orange, and hardwood trees. But the

Acehnese environmental activist Rudi Putra.

community, whose residents are among the millions of Sumatrans dependent on the Leuser for water and food, continues to suffer the fallout of the industry. Flash floods were more frequent thanks to erosion, and incursions by unwanted wildlife had become ever more commonplace. Before the palm oil companies moved in, Ngatimen said, "there used to be all sorts of birds. Now you have to travel very far into the mountains to hear anything."

Putra was determined that I do just that, in order to spend some time in a part of the Leuser known as Ketambe ("the most beautiful place in the world," he'd said by email), where a thirty-year-old research station plays host to scientists studying the area's rich biodiversity. While much of the surrounding region was logged under Suharto, the remote heart of Ketambe remains blessedly untouched. After a forty-minute flight south from the provincial capital of Banda Aceh (site of the devastating 2004 tsunami), the Leuser rippling out beneath us in what looked like green waves, we touched down in a valley cradling a scattering of villages and transferred to a jeep. Wending around twisty roads, we passed mats of brown candlenuts drying in the sun and clutches of little girls marching off to school in long sleeved shirts and matching headscarves—evidence of the strong Muslim tradition alive here on the tip of the island.

A dugout canoe spirited us across a rushing river, depositing us in the sand, and we made our way into the near-dark of the forest. Stepping over fallen logs and leaves in various stages of decay, we navigated through a thicket of trunks with diameters ranging from an eighth of an inch to six feet. A century-old strangling fig, its individual roots intermingling into an eventual whole, reached heroically for the sun. At one point we spotted an orangutan chilling out some seventy feet above us. After staring us down for a good ten minutes, she reached out a fuzzy orange arm, the skinny trunk supporting her bending cartoon-like under her weight, and swung to grasp an adjacent branch. Buzz of cicada, trickle of water over stones. Trill. Chirp. Squawk. The place was a riot of life, black marble centipedes here, saffron butterflies there. (And leeches all over us.)

A huge, gray pheasant with a long tail and little blue head—a great argus—clambered through the underbrush, followed later by a tiny, gray-stomached Horsfield's babbler. From far above came the loud whistle of the Asian fairy-bluebird and the rapid *tu-trruk, ku-trruk* of the diminutive black-eared barbet. A cream-vented bulbul flaunted its fabulous green-yellow wings. But it was a rhythmic, helicopter-like whooshing that stopped us in our tracks. Raising our eyes to the canopy, we caught a pair of wreathed hornbills flap in and out of sight. At another point, two wrinkled hornbills put on a show, one feeding figs to the other, the duo flying off in quick succession. With black bodies and frilled white skirts, they evoked cocktail waitresses in the sky. The elusive helmeted hornbill failed to materialize—not surprisingly, given its diminished numbers—as did its casqued cousin, the rhinoceros hornbill. Though the latter's "ivory" is hollow, the bird also has fallen victim to the poachers, many of whom mistake it for their helmeted prey; Indonesian populations of rhinoceros hornbill are estimated to have dropped to fewer than three thousand. I'd been on hikes in the tropical forests of Kenya, Ecuador, and elsewhere, but this extended stay deep, deep inside such a pristine expanse—we spent a few nights in the researchers' cabins—had been like passing through a portal into some fairytale world.

That orangutan encounter had brought painfully home just how costly the loss of these rainforests would be. The primates— orangutan means "people of the forest" in Malay—live only in

Southeast Asia: here on Sumatra and in the rainforests of neighboring Borneo. In 2008, the Sumatran orangutan was declared critically endangered by the IUCN. Today, its numbers are down to just 14,000, 85 percent of whom make the Leuser their home. The orangutans of Borneo, where the forest has shrunk by 55 percent in just two decades, were deemed critically endangered in 2016. (In 2017, scientists determined that some of the orangutans of North Sumatra in fact comprise a separate species. With an estimated population of fewer than eight hundred, the Tapanuli orangutan, *Pongo tapanuliensis*, is the rarest great ape species in the world.)

The orangutans that remain are fortunate to have an advocate as single-mindedly devoted to them as Rudi Putra is to his rhinos. Hard-drinking, hard-smoking, and as foul-mouthed as Putra is polite, Ian Singleton finishes most of his long days holding rowdy court on the back patio of Roland's, a funky German joint that serves as watering hole to the motley crew of expats who call Medan their home. The capital of North Sumatra province, just southeast of Aceh, the city is a rough-and-tumble sort of place, as well known for its drugs and wildlife trafficking as it is for being a center of the palm oil trade. (As our plane descended over it, the pilot had informed the passengers on my cheap regional flight that the possession of illegal narcotics would be subject to penalty by death.) Having grown up in northern England, Singleton worked as a zookeeper before relocating to Sumatra twenty-five years ago to pursue fieldwork for his PhD in anthropology. He married an Indonesian woman and has lived here ever since. Today, the father of two heads up the Sumatran Orangutan Conservation Programme, or SOCP, which he co-founded in 2001.

He'd agreed to my request for a meeting somewhat grudgingly. Between managing a staff of 120 and juggling the endless administrative tasks involved in trying to save a species on the brink of extinction, Singleton doesn't have a whole lot of time for strangers. He also, having spent two years living in the swamps on the western coast of Sumatra surrounded only by wild orangutans, gives the impression of a man for whom my type of hominoid is perhaps not his favorite type. Still, he welcomed me into SOCP headquarters, a residential building in a quiet-ish section of the city, where the utilitarian décor echoed Singleton's own can't-be-bothered aesthetic: The boyish fifty-five-year-old tends toward cargo shorts and rubber clogs, or, for the office, dark jeans and baggy soccer shirts.

Grabbing a can of instant Nescafé and a jar of nondairy creamer from his mail cubby, he led me up a flight of stairs to his office, a tsunami of notebooks, papers, and reports, with orangutan photos in cheap frames nailed randomly to the walls. He opened his MacBook Air and began walking me through the PowerPoint presentation that he uses in the fundraising efforts he wishes he didn't have to make. Beginning with a quick history of the palm oil industry, he explained how the region's orangutans were being forced from their jungle homes by the ever-multiplying plantations. The primates might try to move to a nearby forest, he said, only to find it already full of other orangutans. With only so much food to go around, many end up dying of starvation. A mother and her infant might wander into a village or plantation in search of something to eat and be shot or clubbed to death. Orphaned infants often were captured and sold as pets. One of his slides showed two SOCP staffers armed with tranquilizer guns and a net, rescuing a marooned and starving animal from the top of a tree surrounded by an endless sea of oil palm. Such refugees get transported to viable patches of forest elsewhere on the island, or, if they've been wounded or orphaned, taken to SOCP's rehabilitation center, known as "the quarantine," located an hour's drive outside of Medan.

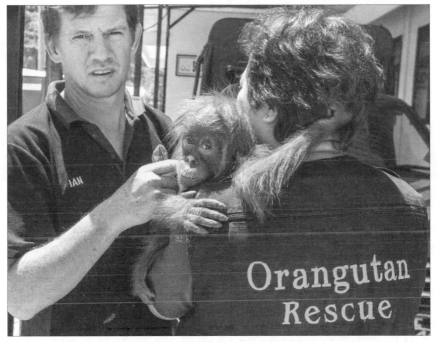

British primatologist Ian Singleton with a young Sumatran orangutan.

Singleton clicked to a slide featuring what looked remarkably like a small child, with tiny fingers grasping the bars of a cage. (Orangutans, which are considered the most intelligent of all primates, share roughly 97 percent of humans' DNA.) "People go, 'Oh, the poor thing,'" he said, "but these are the lucky ones. All the other ones are dead."

His phone buzzed with incoming texts and WhatsApp messages, and the landline rang incessantly. Singleton fielded the calls—and the repeated interruptions by staffers poking their heads in the door—in English, Bahasa, or the local Karonese dialect, as the situation required. He clicked to an aerial shot of the Tripa Swamp, a waterlogged expanse on the west coast of the island, and launched into what I would come to recognize as a typical Singleton rant: "Tripa is classic of Indonesia," he began. "You've got a peat swamp area, it's primary forest, loads of biodiversity. It's full of fish, and it's full of water. And you've got a few local communities that traditionally catch those fish, and it provides most of their protein. All the water supply for the local community is from that water table. And then you've got some people who actually consider they own bits of the land, because it was cleared by their great-grandparents, but they don't have any paperwork. So you get a company then from Jakarta or somewhere who comes in and just evicts everybody. 'Fuck off.' 'Hey, that's my land!' 'Paperwork? Sorry, mate.' So they're kicked off. And then the company offers such shitty wages that none of these people want to work for it. They don't like the company anyway; they've just been evicted. So then they get labor cheaper from offshore islands. And they come and live in shitty conditions. Then the company chops all the forest down. So you annihilate everything that lives there, including ants and termites and funguses. Incinerate the whole fucking thing. Then you dig canals, because in order to grow oil palm you need at least a meter of dry peat to plant the thing. And then the river levels go down, the fisheries disappear, so you've got all these people who used to make their living, and water supplies, and protein source, all of a sudden have got fuck-all. And they're surrounded by plantations. So even if they did have any money, they can't grow any vegetables or fruits. Because it's all—there's one village in Tripa where they asked for a hectare to build a graveyard, and they were refused. So they've got nothing. They're totally screwed. Their water supply's gone, their livelihood's gone, their protein is gone. They don't even want to work there. They've got *nothing*. And then, some company—or some guy in Jakarta who's probably never even been there—his bank account is going up and up and up and up, for twenty-five or thirty years."

(Many of the richest Indonesians identified last year by *Forbes* do trace their fortunes to the palm oil industry. A hotelier friend stationed in Jakarta from 2012 to 2015 recalled for me how the Ferraris, Bentleys, and Rolls-Royces would line up in front of the high-end bars and nightclubs every Friday and Saturday night. "You've seen *Crazy Rich Asians*, right?" he asked. "That's the tip of the iceberg. That absolutely exists." The fortunes of the "twenty or thirty families" that rule Indonesia today, he added, are all based on the country's natural resources, oil palm prominent among them. "Some of the houses in Jakarta, you could pick them up and stick them in the Hamptons, and they would absolutely not look out of place.")

Singleton and I set out for "the quarantine" the following morning. He was grumpy—we'd had a late night at Roland's—and, maneuvering his lime-green Jeep through the anarchy that is Medan rush-hour traffic, responded tersely to my questions. I took the hint and shut up, taking in the view as we wound around and around the hilly terrain, cruising through towns of open-fronted, two-story cement buildings hung with signs for Panamas, Lucky Strikes, and a million other brands of cigarette. Braking and accelerating like a pro, he tooted at other drivers to nudge them out of the way and routinely darted back to our lane just as smoke-belching trucks barreled straight for us. "You just have to expect everybody to do everything all the time," he said. "That's the irony. If you overtake somebody on a blind corner and see three trucks coming toward you, you're expecting to see them, and they're expecting to see you. So it kind of works."

We stopped to pick up enormous bags of bok choy, sweet potatoes, corn, carrots, green pepper, star fruit, cauliflower, and cabbage for the orangutans—"they like everything we do"—and proceeded down a long, curving driveway to arrive at what looked like a wilderness camp, with wooden outbuildings scattered among dense foliage. Securing a celery-colored surgical mask over my face, I followed an American staffer named Jennifer Draiss along a winding path in the direction of an insistent screeching sound. We paused before a caged-in expanse, and she pointed out a ten-month-old orangutan named Didi, all scraggly fur and huge eyes, who was vehemently protesting an apparent fruit theft by her cage-mate, a fifteen-month-old named Deka. Didi had arrived at the sanctuary four months earlier, Draiss said, her mother having succumbed to an attack by villagers, likely involving clubs, machetes, or air rifles. Deka had been delivered to the quarantine a few weeks after that, having been confiscated from an illegal-pet trader. Local farmers often kill the mother orangutans in order to obtain the infants, which they sell to traders just as the hornbill poachers do. In this case, Didi had taken a pellet to the skull in the process. Orangutans in the wild will stay with their mothers

On the Indonesian island of Sumatra, rescuing an orangutan marooned in a sea of oil palm.

for up to twelve years, learning how to build nests, find food, and otherwise survive in the jungle. Being abandoned at this young age, she said, had left the pair virtually helpless.

"She's getting fat," Draiss noted of a diapered infant as we stepped into the baby house. "If you give her water instead of milk, she'll bite you." Given that an orangutan in the wild will nurse for as many as seven years, you couldn't really blame the poor creature. Across the room, a fifteen-month-old named Dina, suffering from meningitis and on twenty-four-hour care, lay on her back surrounded by pillows. She tried to sit herself up but flopped clumsily down on her face. The babies get taken for walks every day, Draiss explained, so that they learn to recognize the sounds of the jungle.

She led me to the clinic, the first stop for any orangutan arriving at the quarantine. (There were a total of forty-eight in residence when I visited.) Here, the animals get x-rayed to check for tuberculosis, have their blood screened for hepatitis and herpes, and undergo a dental check before being tattooed for identification purposes. After a designated quarantine period, healthy residents can then move on to the "socialization cages," where they meet other orangutans, often for the first time since having lost their moms and been captured. Set seven feet off the ground and measuring about four hundred square feet, these cages were where the real action took place. The orangutans swung from strips of rubber suspended like vines from the ceiling and hung upside down, munching on bananas and dropping the peels to the floor. They hooted and squealed, offering us gleeful, teeth-baring kiss-offs. "This is where they really learn how to defend their food, who's dominant, who's subordinate," Draiss said. "In the baby house, they cry when we go by. Once they're in socialization, they want nothing to do with us anymore. It's nice to see."

She led me on a narrow path through the trees, our shoes crunching

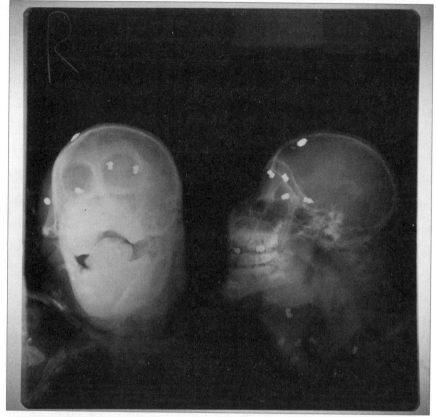

X-ray of Leuser, an orangutan shot sixty-two times on the island of Sumatra.

over a carpet of dried leaves, and to a clearing where the staff had built what they call the "forest school." Three young orangutans, each about twenty pounds, were swinging from the branches high above, and plucking mangoes from wire baskets that the staffers had strung up using pulleys. They were learning to find food and to make nests from leaves, while also building muscle tone to get fit for eventual life in the wild. Once they're deemed ready, the orangutans are taken to one of two SOCP reintroduction sites, one located south of Medan, in Jambi province, and the other a twelve-hour drive north, in a place called Jantho.

We walked by a cage that appeared empty, until I spotted a heap of orange fur huddled in the far corner. That was Leuser, Draiss said, an adult male originally brought to the center back in 2004, when he was four or five years old. He'd been confiscated from a military man by Singleton's team and the wildlife authorities and, after months of reintroduction training, had been returned to the wild in Jambi. Two years after that, he'd been found by the wildlife authorities after reports that an orangutan had been in "an incident" at the edge of the forest. Leuser had sixty-two air-rifle pellets in his body, including in both of his eyes. The vets had managed to remove fourteen of the pellets, but Leuser had been totally blind ever since. In 2010, he was introduced to a female orangutan named Gober, also blind, who had eventually given birth to twins, a male and a female, both of whom had eventually been sent off to live on their own. In 2015, an Indonesian eye specialist had successfully

One of the twins sired by Leuser and Gober, blind Sumatran orangutans.

operated on Gober's cataracts, and she now lives with the twins in the forest.

Singleton caught up to us at the cage of a large adult male named Jinto. "That's a nasty little hole on you, isn't it?" he said to the animal, in the manner of a father addressing his toddler. He moved in to inspect a machete gash in the orangutan's shoulder. "You got a stomach on you, too, mate," he said, patting his own midsection. Jinto had been released last year, Singleton explained, but they'd found him again three weeks ago with this fresh wound. "Why are you coughing like that, mate?" he asked, turning his attention back to Jinto.

"My work is mostly crap," Singleton told me a few nights later over more beers at Roland's. "It's mostly putting out fires, dealing with government, meetings, sitting in the airport. The journalists, the government people, the funders who want all my time?" he shook his head. "But occasionally I get up to Jantho and I see some of these orangutans in the trees. And they're behaving like wild orangutans, and they look down at me, and they're not interested, and it's fantastic."

Earlier that afternoon, touring me around the open-air reserve he's begun building for primates unable ever to be released on account of health or disability reasons, Singleton had pointed to a patch of sunflowers that he'd recently planted himself. "The orangutans really like the seeds," he said, "and you can only buy the salted ones in town."

My last day in the Leuser, Putra and I set out in the morning to investigate some smoke that he'd noticed the previous afternoon, and which he suspected was rising from within the national park. After an hour of switching back over tawny dirt roads and, as seems inevitable, getting lost amid the maze that is an oil-palm plantation (the smoke rose from its far side), Putra punched some numbers into his Nokia and a man named Pranyoga materialized on a red motorbike. "He is the best of my spies," Putra said. "I call him 'the man without afraid.'"

Pranyoga, who grew up nearby in a forest that's since been supplanted by oil palm, had worked with Putra for sixteen years, serving as a liaison to the community and keeping a watchful eye on the comings and goings, often illegal, of the industry. Though he'd had repeated threats to his life, he said he was determined to ensure that his own kids get the chance to appreciate the elephants, sun bears, orangutans, magpies, and hornbills that he remembered from his own childhood.

Our driver followed Pranyoga's bike up, down, and around endless curves until we eventually reached a ridge and found ourselves looking out over a picture painted in black. Trudging through the still-smoldering ash, Putra estimated that the blaze had been set a week earlier—some 150 acres, he later estimated, of secondary lowland forest torched to the ground. The culprits, he imagined, were land-starved locals looking to cultivate rubber, cacao, and oil palm as a source of income. We picked our way around charred stumps and over the mottled beige and cinnamon corpse of a Burmese python. Clomping over brittle twigs and roasted ferns, I noted the weird dissonance of the pleasant, Indian sandalwood–scented air. "The government says it doesn't have the budget," Putra said. "But we can prevent this if they cared about it."

Eventually the dry tree leaves rasped and the air began to fill with smoke. "Jocelyn, look!" Putra called, pointing to a mass of orange growing in the distance. The hungry flames ripped toward us, crackling and popping, as ashes floated down like ebony snowflakes. With smoke filling our lungs and stinging our eyes, we dashed back to the car and, streaked with black, raced away from the heat. On the way out, we passed yet more industrial plantings, right there inside the national park. "I hope this fire will reach the palms," Putra mumbled, to no one in particular.

That fire would have died down soon enough, but similar conflagrations have become a recurring thing on Sumatra, a phenomenon that's taken a horrific toll on human lives. A 2019 study published in the *Proceedings of the National Academy of Sciences* found that early-life exposure to the haze caused by such tropical forest fires results in long-term, irreversible health impacts, and in the economic losses they lead to.

The locals in Aceh didn't have the luxury of worrying about anything so far in the future. They were simply doing what it took to get by. The three poachers I'd met in Tamiang, for example, all of them "ex-combatants" (veterans of the independence movement and, like most of their former comrades-in-arms, unschooled and ill-prepared for formal employment), were likely planning their next three-week foray into the forest in search of the prized helmeted hornbill. They didn't particularly understand the Chinese buyers' obsession with the birds—they'd heard that the foreigners use the casques for jewelry, or as toys for their kids—but they knew that they would find a ready market. "The moment you come down from the mountain," our host had explained, "immediately someone will come and take the casques to Medan." The sale of a single beak, he'd added, would mint enough to feed three families for a month. "Whatever reason they might want them," the gun owner chimed in, "we will sell them. If there was an easier job, especially if it wasn't illegal, of course we would choose that."

In fact, he had shelled out extra money to have his rifle customized to shoot 5.5-millimeter pellets. The standard 4.5 ones, he said, tended not to kill the hornbills on impact, and he and his buddies couldn't stand to see the birds suffer.

6

Caravan Dreams

It was there that the sleight-of-hand lawyers proved that the demands lacked all validity for the simple reason that the banana company did not have, never had had, and never would have any workers in its service because they were all hired on a temporary and occasional basis . . . and by a decision of the court it was established and set down in solemn decrees that the workers did not exist.

—Gabriel García Márquez, *One Hundred Years of Solitude*

WALTER BANEGAS was fifteen minutes from finishing his shift when he made his way over to the day's final row of oil palms, the one abutting the plantation's main road. It was just before 2 p.m. on a Thursday, and having clocked in at six, the thirty-four-year-old was looking forward to peeling off his sodden shirt and ditching the muggy rubber boots. But as he maneuvered his harvesting *malayo* to dislodge a ripe bunch of fruit, his left foot slipped in the wet grass, sending the twenty-foot aluminum pole listing toward the power line that runs the length of the road. "From there," he told me, "I don't remember anything else."

His co-workers sure do. They would later describe to Banegas how his body had exploded into a ball of fire, as if out of some superhero movie—only no muscle-bound savior charged from the smoking wreckage. Instead, Banegas' body lay crumpled on the ground. Now, sitting across from me on one of the plastic chairs that we'd dragged inside to escape the punishing Honduran sun, he was recounting how the rest of that fateful day four years ago had played out.

His fellow harvesters had fanned out in all directions, running frantically through the rows of palms in hopes of finding someone with a vehicle to get Banegas out of there. The plantation, run by a Honduran company called Agroguay, has neither medical clinic, staff doctor, nor ambulance—despite the fact that its workforce numbers in the hundreds. Forty minutes after the accident, when they had finally located a car, his colleagues hustled Banegas into the backseat, and the driver raced toward the nearest "social security," as the country's government-run clinics are known. But it was another hour before they reached the small city of El Progreso, where clinic staff took one look at him and shook their heads: he would need to see the pros in San Pedro Sula. Three more hours passed by the time they got him to the hospital in the city (capital of the Cortés department, here in the country's northwest), with treatment in the interim limited to the timid application of water-dampened rags on the charred ruin that had been his flesh. Six days in the intensive care unit had been followed by another two months and four days in the hospital, where burn specialists performed skin grafts every third day.

With his hair buzzed short on the sides and a still-strong upper body, Banegas brought to mind a high-school lacrosse player, if a particularly subdued one. An insistent rooster crowed, and his wife

Electrocution victim Walter Banegas at his home in northern Honduras.

and five-year-old daughter sat silently by as the young father considered the impact that the accident had had on his life and that of his family. Though a fourteen-year veteran of Agroguay, Banegas was still considered a "temporary worker" at the time of the incident, so he knew that his medical coverage would be limited. Today, what remains of his lower left arm consists of a toddler-sized wrist webbed in pink and purple scar tissue, and a puffed-up and all but lifeless hand. After the doctors had amputated his right arm just below the elbow, they'd grafted skin from other parts of Banegas' body in order to save the left one—and spare him a fate as a double amputee. "I have burns all over my body," he said, using his left elbow to push aside his shirt and pants, revealing scars plastered across his legs, shoulders, and left foot, now with just two toes. "All these are skin grafts," he said. "Skin graft. Skin graft. That's a burn. That's another burn." A giant scar slashed across his stomach, like the memory of a botched cesarean. Four years later, he said, the pain was still naggingly present. "Mostly it's the burns. And it itches."

Since the introduction of the sickle-fitted harvesting pole, electrocution has joined the long list of occupational hazards common to the global palm oil industry. (A handful of other Hondurans, including one of Banegas' colleagues, have recently suffered similar accidents, some fatal, as have workers in Southeast Asia. In 2019, two laborers commuting in a flatbed truck from an oil-palm plantation in Uganda were killed when their sickles hit an electric wire.) Over the course of my reporting, I heard the same stories over and over—of long hours, low wages, dismal housing, non-existent health benefits, exposure to dangerous chemicals, inadequate safety gear, sexual abuse, and fatal snakebites—from plantation workers across Africa, Southeast Asia, and Latin America. In many places, the labor situation appeared little improved from the days of the contract coolies in Malaya and the conscription-by-*chicotte* of the Belgian Congo. (Given the toxic chemicals involved in today's industry, in fact, conditions may be worse.) Here in Honduras, where oil-palm plantations now cover nearly 500,000 acres—palm oil is the country's fourth-largest export—oppression at the hands of the plantation master goes a long way back.

In the early years of the twentieth century, American-owned companies took over large expanses of fertile land in Honduras and converted them to fruit plantations, mostly growing bananas and pineapples. By the 1920s, the United Fruit Company, founded in 1899 by a Boston importer and a wealthy young New Yorker who'd been growing bananas in Costa Rica, was overseeing an empire that extended the length of Central America. In 1911, a New Orleans–based banana man named Samuel Zemurray—his Cuyamel Fruit Company would later be absorbed by United Fruit, which he would go on to lead—became so incensed with the Honduran government for failing to deliver the land concessions and tax breaks he'd requested that he hired a pair of American mercenaries to overthrow it. (The American author O. Henry coined the term *banana republic* to describe the fictional country of Anchuria, based on his own experience living in a Honduras ruled by foreign fruit companies.) Among Hondurans, United Fruit came to be known as El Pulpo, or "The Octopus," for the way its grip extended to everything from the country's transport infrastructure to its telecommunications systems.

By the 1930s, United Fruit had taken control of some 3.5 million acres across Central America and the Caribbean. In neighboring Guatemala, it owned more land than any other interest. Demure guest it was not. "Through bribery, fraud, chicanery, strong-arm tactics, extortion, tax evasion, and subversion [United Fruit] grew to be a swaggering behemoth," Stephen Schlesinger and Stephen Kinzer write in their 1983 book *Bitter Fruit*. Thomas McCann, a former vice president of public relations for the company, later admitted that the United States had trained an army of mercenaries on its Central American plantations. (The Chiquita name and jingle, along with the cartoon figure Senorita Chiquita Banana, modeled on the sexy Brazilian performer Carmen Miranda, were introduced in 1944 as a way of softening the brand's image back home.) In 1952, after Guatemala's democratically elected president attempted to turn over a quarter-million acres of the company's fallow land to poor farmers, Zemurray enlisted a PR man named Edward Bernays to navigate the conflict. A nephew of Sigmund Freud, Bernays had, since the 1928 publication of his book *Propaganda*, earned a reputation as "the father of public relations." Invoking the usual "communist terror," he teamed with the CIA to pull off yet another Central American coup. A decade later, United Fruit would again conspire with the CIA— this time to abet what would become the debacle at the Bay of Pigs.

Beginning in the 1940s, in part as a hedge against the Panama blight then ravaging its banana crop, United Fruit undertook modest oil-palm plantings in Honduras and Costa Rica. (Back in the mid-twenties, the company had built research centers at La Lima and Lancetilla, near its headquarters in the Honduran coastal town of Tela, importing genetic material from oil-palm operations in Africa and Southeast Asia.) By the 1960s, it had established a large collection of *Elaeis oleifera*, an obscure oil-palm species native to South and Central America with which it would eventually develop a hybrid using Africa's *Elaeis guineensis*. Such "American oil palm" is now grown commercially across Central and South America. Also in the 1960s, United Fruit acquired a Costa Rica–based vegetable-oil concern, called the Numar Company, and founded Grupo Numar, specializing in oils and fats. In 1995—United Fruit having by then changed its name to Chiquita Brands International—it sold off Grupo Numar, which merged with another operation to form Grupo Jaremar, the parent of Walter Banegas employer Agroguay. One of the largest companies in Honduras, Jaremar today oversees oil-palm plantings covering some 37,000 acres, in addition to running its own mills, refineries, and manufacturing sites. Among its retail products are such regionally adored brands as Clover, Mrs. Pickford's, and Blanquita (margarines and oils); Max Poder (laundry soap); and Riki Tiki (cookies and chips). While most of the company's palm oil gets sold in Central America, some of it ends up in products made by such American companies as Hershey's, PepsiCo, and Entenmann's parent Grupo Bimbo.

Happy to leave charmless San Pedro Sula behind, I set out one Sunday morning in the company of a labor-rights activist to visit a Grupo Jaremar plantation and meet some of its workers. What the city lacks in colonial architecture, it makes up for in twenty-first-century terror, consistently ranking among the most violent urban centers in the world. Stationed in the small lobby of the hotel that I'd chosen for its reputed safety were not one but two automatic-weapon-wielding guards. That morning, we made our way out of downtown and headed southeast, past a series of mini-malls anchored by a McDonald's and a Popeyes, a Burger King and a Wendy's, and eventually merged onto a four-lane highway. Driving back to town one night later that week, this stretch of road would be flanked by workers streaming out from some of the country's hundreds of *maquilas*, or factories, having spent the day bent over sewing machines stitching T-shirts and underwear for the likes of Nike, Under Armour, Dickies, and Champion. Not normally a nervous traveler, I'll confess to having been somewhat frantic during the long minutes we spent stuck in traffic; it would have been easy for someone with a gun to force his way into our car.

If a sense of insecurity pervades the place, it's a situation for which the United States shoulders a fair portion of the blame. During the Reagan years, the U.S. government used Honduras as a base for fighting Nicaragua's leftist Sandinistas, and for training the Salvadoran army during that neighboring country's civil war. In the service of the "war on drugs," we armed the Honduran security forces, which further helped to militarize society here. More recently, United States–led interdiction efforts in Mexico and the Caribbean have led narcotics traffickers to begin funneling their cocaine shipments through Honduras and neighboring Guatemala and El Salvador. A 2017 study published in *Environmental Research Letters* found that roughly 86 percent of the cocaine trafficked globally now moves through these Central American countries. The six or so billion dollars in illegal profits made every year need to be laundered somewhere, and the criminals involved have found that oil-palm plantations work quite nicely. Much of the contraband passes through northern Guatemala's Petén region, where oil palm also has supplanted smallholder farms.

In the early 1970s, the Honduran government enacted a national reform law that saw the redistribution of land, including the rich terrain here in Cortés and in the neighboring Aguán Valley, to collectives run by peasant farmers. Settlers to the region were provided seeds, plantings, fertilizer, and credit to grow bananas, citrus, and oil palm. Two decades later, however, a new administration passed laws allowing for the land owned by such collectives to be broken up and privately sold. Large expanses fell into the hands of a few wealthy businessmen, among them René Morales Carazo, owner of Grupo Jaremar, and a man named Miguel Facussé, whose Dinant Corporation dominates the country's palm oil industry. (Dinant controls some 29 percent of production, as opposed to Jaremar's 24 percent.) After coming to power in 2006, Honduran president Manuel Zelaya began working to resolve the unrest that had resulted from the land disputes, but he was ousted in a military coup in 2009. Displaced farmers in the region have been in bitter conflict with the oil-palm companies ever since. Between 2009 and 2012, fifty-three murders of peasants in the region were attributed to guards and mercenaries hired by the large oil-palm growers, often acting with the help of the state's police and military forces.

Roughly an hour after leaving the city, we reached El Progreso, where Banegas had been taken to the clinic, and another twenty minutes found us deep in banana country, the floppy-leafed plants lining either side of the dirt road. The picturesque Mico Quemado, or "Burned Monkey," mountains rose in the east. Soon the view changed from bananas to oil palms, expanding out in neat rows.

After owning this fertile expanse for nearly a century, Chiquita had, after the devastation of 1998's Hurricane Mitch, which killed some 6,500 Hondurans, sold much of the land to Grupo Jaremar. A worker pedaled by holding his *malayo* aloft—it extended several feet out in front of and behind him—and a pair of women approached toting either end of a sack of loose oil-palm fruits. This would have been in service of the job, as Hondurans don't use palm oil at home. In fact, the workers I spoke with said they had no idea where the stuff they harvested even ends up.

We pulled up to the wood-slat house of a fifty-four-year-old I'll call Carlos and settled in on his cement patio. A thoughtful man with brown eyes and baggy pants tucked into knee-high rubber boots, he explained that he had begun working for Chiquita in 1995, watering banana plants and harvesting the fruit. Back then, he'd had a contract and was making 250 lempiras ($10) a week, enough to support his wife and two children. After Mitch, Carlos had signed on with Jaremar, but six months into the job, once the oil-palm seedlings were all in the ground, the company had dropped his pay to 200 lempiras ($8.14) a week. The permanent work staff, which under Chiquita had numbered some three hundred, was reduced to seventy, and within three years, between reprisals and firings, only thirteen permanent workers remained. Finally, they were down to just six. Now employees were hired on a series of temporary contracts (in violation of Honduran law), and the company provided neither education vouchers nor sick leave nor vacation days. It took monthly deductions of 370 lempiras ($15) for healthcare, but when the employees went to access services, often they were told they were unavailable. Carlos can no longer afford to pay for his children to go to school. "The company says it gives scholarships," he said, "but only to workers who are friendly with management. I want to die when my daughter says that she wants to study more."

We headed back toward the main road and stopped alongside a handful of young men resting in the shade of the palms. They agreed to chat on the condition that I not divulge their names. (We had made it a point to visit on a Sunday so as to avoid contending with the company's security guards.) A thirty-two-year-old in a baseball cap, a fourteen-year veteran of the company, yanked off one of his company-issued rubber boots to show us how flimsy they were. The snakes that lurked in the grass, he said, bite right through them with their sharp fangs. They include the *barba amarilla*, or "yellow beard," one of the most dangerous snakes in the Western hemisphere. Part of the problem, said the guys, was that the bosses were too cheap to keep the plantations clean. "When there's production, they hire a lot of people to harvest," explained one. "But when it's done, they fire a lot of people, even though there's work to be done cleaning. It's a mess between the trees."

The chemicals handled by Agroguay workers and that accumulate in the undergrowth are known to be extremely dangerous. Among them is an insecticide called Lorsban, the commercial name for chlorpyrifos. An organophosphate belonging to the same class of chemicals as those developed by the Nazis for use as a nerve gas, chlorpyrifos has been linked to a range of medical conditions, including brain damage in children, Parkinson's disease, and cancer. In January of 2020, the European Food Safety Authority, citing safety concerns, announced that it would not renew the permit for its use. The company also uses glyphosate, the key ingredient in the Bayer (formerly Monsanto) herbicide Roundup. In 2015, the World Health Organization (WHO) declared glyphosate a probable human carcinogen, linking it to non-Hodgkin's lymphoma and to DNA damage, and Bayer has since settled a class-action lawsuit related to such claims. Though Agroguay workers charged with applying fertilizers and pesticides were provided gloves and face masks, said the workers, the masks fogged up in the heat, rendering them useless. "We have told the company, 'You put the equipment on. You'll see that you cannot.'"

The female employees whose job it is to pick up loose fruits from the overgrowth complained of vision problems and of itchy skin conditions on their hands and arms. And though the WHO recommends that workers dealing with toxic chemicals wash immediately afterward to prevent "hazardous contamination," there are no bathing facilities on the Agroguay plantation. It takes most of the company's workers at least thirty minutes to get home by bicycle.

In the film scene, a worker wearing a headscarf and a backpack sprayer aims its three-foot hose toward the ground surrounding rows of oil-palm plants on a plantation in North Sumatra. Drawing her free hand across her throat as if to slit it, she swoons dramatically backward. "The poison stinks!" she says, hamming it up for a pair of nervously giggling colleagues. "That's Gramoxone. It makes me want to throw up. I have a rash on my buttocks from the Gramoxone," she goes on in an exaggeratedly cheery, TV-commercial voice. "The chemical is ruining my eyes." She has tried going to the company doctor for medicine, she tells her friends, but he told her, "Your eyes are fine," and refused her request for drops. "So I asked my friends, 'Don't your eyes hurt?' And they said, 'There's nothing worse!' They said, 'Our backs itch too!' 'So let's ask for plastic suits,' I said. 'Even our hair is falling out because of Gramoxone!'"

The deeply disturbing footage, from a 2003 documentary called *The Globalization Tapes*, was shot on Adrien Hallet's original Socfin estate, near that erstwhile corpse-clogged river. Members of Serbuk, the Independent Plantation Workers' Union of Sumatra, wrote and directed the film, with the help of Joshua Oppenheimer, the American director behind 2012's Oscar-nominated *The Act of Killing*, about Indonesia's 1965 genocide. (The latter film grew out of this original project.) Gramoxone is the commercial name for an agricultural chemical called paraquat. Despite having been linked to kidney, lung, and liver damage, and, like chlorpyrifos, to Parkinson's disease—it's banned in at least forty-six countries—paraquat continues to be used on oil-palm plantations throughout the world. It is generally women like the ones in the film, often wearing little or no protective gear, who are charged with handling paraquat, which can be fatal if inhaled, ingested, or absorbed through the skin. (Fans of the filmmakers Ethan and Joel Coen might recall the scene from their 1998 hit *The Big Lebowski*, in which Jeff Bridges' "Dude" spits out the ultimate diss—*"human paraquat"*—at an adversary. In the 1970s, the U.S. government had the chemical sprayed on marijuana plants across Mexico.)

Though most palm oil companies have vowed to stop using paraquat, the evidence suggests that few have actually done so. In 2008, for instance, industry giant Wilmar said that it would require its suppliers to phase out paraquat by the end of 2015. But a 2016 report by Amnesty International documented its use by the company's Indonesia-based suppliers. (Wilmar, among whose major shareholders is the American commodities giant ADM, is the largest agribusiness in Asia. It owns a network of plantations, mills, and refineries around the world.) In a 2014 statement to the Overseas Private Investment Corporation, or OPIC, which provided it with a $20 million loan, Honduras's Grupo Jaremar wrote that "Some paraquat containing pesticides are currently used but will be phased out."

En route to work on an oil-palm plantation in Sinoe County, Liberia.

Where paraquat *has* been eliminated, it generally has been replaced by glyphosate.

The Amnesty researchers spoke with Indonesian mothers who applied pesticides without wearing protective gear and who were forced to wake at 4 a.m. and feed their children before traveling as long as two hours to plantations. Returning home in the evenings, they often had to choose between using their limited water supply to rinse off the dangerous chemicals or to cook dinner for their families. A woman named Yohanna said that she had been splashed in the face with Gramoxone. "I can't see through the eye," she told the researchers. "I get headaches." In addition, as if out of some nightmare Henrietta Lacks scenario, Wilmar subsidiaries and suppliers were routinely testing their female employees for chemical exposure but then failing to provide them with the results. Those whose tests revealed anomalies were simply told that there was a problem and were moved to other tasks. "Every six months we are checked," said a woman who had been spraying pesticides for five years. "Our blood is taken. But we are never told the results. So if in our body there is an illness, we don't know."

A 2015 report by an Indonesian non-governmental organization called Sawit Watch found that female workers who spent hours each day spraying fertilizers, pesticides, and herbicides on the oil-palm plantations of the Singapore-based company Musim Mas were given milk or pudding intended to offset the effects of the exposure. "They told us that it was to help us detoxify," one laborer told the researchers behind the report. ("My friends asked, 'Do they give you milk?'" says the woman from the scene in *The Globalization Tapes*. "'Once every three months.' So I asked the boss, 'Why do you give us milk?' 'To get the poison out of your chest,' he said, 'and so you don't get dizzy,' he said. So he knows we're being poisoned!")

Female plantation workers across Africa, Latin America, and Southeast Asia also have reported rape and sexual abuse on oil-palm plantations. In a November 2020 investigation published by the Associated Press, a sixteen-year-old laborer on a Sumatran plantation described how she had been raped multiple times, and eventually impregnated, by a supervisor old enough to be her grandfather. "He threatened to kill me," she told the reporters. "He threatened to kill my whole family." Another worker said that she had lost two babies in the third trimester as a result of carrying heavy loads of fertilizer. She'd kept her pregnancies a secret for fear that divulging them would get her fired.

A 2015 article in the *Wall Street Journal* told the story of a twenty-two-year-old Bangladeshi man named Mohammed Rubel who had found himself a virtual slave on an oil-palm plantation in Malaysia. Rubel told the reporter that a stranger had shown up in his home village and talked excitedly about the lucrative jobs to be had on plantations in the neighboring country. His family had scraped together $2,000 to pay the recruiter, and Rubel had boarded a forty-foot fishing boat along with some two hundred other Bangladeshis and ethnic Rohingya Muslims from Myanmar. Over the course of a three-week journey, he said, the armed men manning the flimsy vessel rationed food and water to such an extent that dozens of the would-be laborers died. (To prevent the corpses from ballooning up and floating, the traffickers slit their stomachs before dumping them overboard.) Arriving in southern Thailand, the men were marched into crowded camps surrounded by barbed wire. The traffickers then demanded that they place calls to their parents, and, beating them so that their screams would be audible on the other end of the line, forced the young men to tell their parents that the captors "would bury [them] in the jungle" if they didn't send more money. At the camp, Rubel said, he saw dozens more migrants die from disease, exhaustion, and beatings.

After being forced to march for days through the jungle, the captives finally arrived in southern Malaysia, where they were put to work on an oil-palm plantation run by Felda Global Ventures Holdings Berhad (FGV), the commercial arm of the Federal Land Development Authority, or FELDA, introduced in Chapter 4. Today, the state-owned company, which ranks among the largest producers of palm oil in the world, employs some thirty thousand migrants on its Malaysian plantations. Rubel told the reporter that he had worked on FGV's plantations seven days a week without being paid a single ringgit. Another Bangladeshi said that he'd been shunted among three labor contractors over the course of six months in Malaysia, also without receiving any pay. "They buy and sell us like cattle," he said.

In the five years since the *Journal* published its exposé, FGV has done little, if anything, to address the abuses. In 2019, in fact, a group of United States–based labor, environmental, and justice organizations filed a complaint with U.S. Customs and Border Protection seeking to stop the importation of palm oil products produced by the company. (Under the Tariff Act of 1930, U.S. Customs is required to deny entry to goods arriving at our ports if there's reasonable cause to believe they contain materials made with forced labor.) The complaint cited evidence that FGV had exploited victims of trafficking and failed to provide them with adequate food and housing. Migrants on its plantations had been forced to give up their passports, forbidden from leaving, and compelled to sign contracts in languages they didn't understand. Palm oil produced by FGV gets traded by such American companies as Cargill and ADM, and it ends up in the products of Nestlé, Colgate-Palmolive, Johnson & Johnson, Procter & Gamble, Kellogg's, Mars, PepsiCo, and L'Oréal. In September 2020, after an investigation by the Associated Press documented ongoing abuses on FGV plantations, Customs and Border Protection's Office of Trade announced that shipments of palm oil products or derivatives traced to FGV would in fact be detained at U.S. ports.

Next door in Indonesia, plantations that supply Wilmar have been found guilty of employing children as young as eight years old. Many owners establish quotas—for fruit to be collected, for instance, or pesticides applied—and the pressure to meet them has resulted in large numbers of "casual laborers," namely women and children, who work for no pay at all. On some Wilmar plantations, elementary school–aged kids were seen carrying heavy loads of fruit and being exposed to some of the chemicals mentioned earlier. One twelve-year-old told the researchers that he helped his father six days a week. "I pick up the loose fruits. It is not tough to pick up the fruit but there is a small worm that bites me. I put the fruits into the sack and carry it to the collection point. . . . I don't wear gloves and it hurts to pick them up. I don't wear boots, I wear sandals. I work when it is raining, [and] it is slippery. I slip while carrying the sack. I have fallen sometimes."

Reaching some of the plantations formerly run by William Lever entails catching a flight from the Congolese capital of Kinshasa to the provincial capital of Mbandaka and then either driving for several days through the rainforest or traveling 745 miles up the Congo River by boat. Which may explain why the company now overseeing the concessions is able to conduct business in the way that it does. In 2008, Unilever sold three plantations belonging to its Plantations et Huileries du Congo, or PHC, to a Canadian company called Feronia. Two of them are among the five circles originally leased to Lever in 1911. The plantations combined employ nearly ten thousand laborers.

Workers for PHC routinely handle both chlorpyrifos and glypho-

Child laborer on an oil-palm plantation in Malaysia.

sate. Though most of the world's oil-palm companies hire women to mix and apply their chemicals, in PHC's operations this work tends to be done by men. Of forty-three male employees interviewed by Human Rights Watch in 2019, twenty-seven said that they had become impotent since beginning the job. "Sexually, I feel weak," reported a thirty-two-year-old who'd been applying pesticides for two years. "I don't have the strength to satisfy my wife in bed. It shames me." Another employee, who'd been applying pesticides since 2016, explained, "At first I thought I was the only one with this problem and that I should not talk about it with others, but when I heard others, I opened up about what I had become." (The workers generally used the euphemism "sexual weakness" to refer to their conditions but later clarified that they were having difficulty achieving and maintaining erections.) A systematic review published in the *International Journal of Occupational and Environmental Medicine* in 2019 found that male oil-palm workers in Sabah, Malaysia, were suffering from widespread abnormal sperm.

Like their Honduran and Indonesian counterparts, the Congolese laborers described skin irritation, pustules, itchiness, and blisters resulting from the chemicals that they handled. Others spoke of pain and irritation in their eyes and said that their vision had become blurred since starting the job. "We asked for goggles several times," reported a thirty-three-year-old father of six. "They told us to wait but we haven't gotten anything yet." Of forty-three workers interviewed who handled pesticides, only ten said that they'd received goggles. These employees, too, had had their blood taken by the company but never received the results of the tests.

They likewise complained that their rubber boots ripped easily, leaving them vulnerable to snake and spider bites and to machete and thorn-prick injuries. The gloves provided by the company were made of cloth and leather, which, given the way the fabric absorbs and holds the pesticide close to the skin, may be more hazardous to the workers than if they wore no protection at all. Their company-issued overalls were made of permeable cotton, even though they had been told during training that they needed waterproof overalls. Most traveled a considerable distance to and from their homes, increasing their exposure time to the contaminated clothing and exposing their families to the chemicals. As on the plantations of Grupo Jaremar, there were no shower facilities on site.

PHC, too, relies on temporary contracts and day-labor schemes—as of December 2018, the company had nearly seven thousand day laborers in its employ—enabling it to avoid providing benefits that would be owed to workers with contracts. Many of the Congolese

Father of five and twelve-year veteran of the Congolese company PHC, in front of his home.

said that they'd been employed as day laborers for years at a time—even, in at least one case, for a decade. Most also said that they could afford to eat only once a day. One mother of six, who had been picking up loose fruits for more than six years, said that she earned between 12,000 and 15,000 Congolese francs ($7.30 to $9.10) per month. "We work without boots, without gloves—with our bare hands," she said. "Sometimes the fruits [we have to pick up] fall into cows' or people's excrement."

In his 1967 novel *One Hundred Years of Solitude*, Gabriel García Márquez tells the story of Macondo, whose residents' lives are upended when an American banana company comes to town. Having endured inhumane housing conditions, nonexistent medical services, and payment in the form of Virginia ham rather than in cash, the company's desperate laborers finally decide to strike. The government responds by inviting them to gather in the town square for a dialogue. When some three thousand workers show up, the army proceeds to mow them down with machine guns. Márquez based his scene on an event that had taken place in Ciénaga, a town on the Caribbean coast of his native Colombia, back in 1928. There, workers for United Fruit had staged a strike of their own, prompting the government to likewise call for a gathering in the town's central square and to then sanction a massacre. "I have the honor to report," the U.S. ambassador to Colombia wrote in a telegraph to Washington, "that the Bogotá representative of the United Fruit Company told me yesterday that the total number of strikers killed by the Colombian military exceeded one thousand."

After more than twenty years of earning less than the country's minimum wage, working fourteen-hour days without receiving overtime, and being denied health benefits, workers at Honduras' Grupo

Jaremar also decided to strike. In October of 2017, 104 laborers for Agroguay and 80 from its sister company, Agromeza, announced that they had affiliated to El Sindicato de Trabajadores de la Agro-industria y Similares, or STAS, a union representing workers in the country's banana, melon, sugar, and palm oil sectors. Grupo Jaremar responded by firing eighteen of the workers it suspected of leading the movement, prompting three hundred of their colleagues to walk off the job. The strike lasted five months, during which time, the laborers told me, some of them went for days without eating so as to feed their kids. Some were evicted from their homes.

Wearing dark eyeliner and a rhinestone barrette, a thirty-two-year-old named Yarleni Ortéz Mejía invited me into the cramped concrete house that she shares with her two children. After working for Agroguay for seven years, she said, collecting loose fruits from 6 a.m. to 2 p.m. Monday to Friday, and from 6 to 10 on Saturdays, she had been making just 300 lempiras ($12) a week. She'd been reluctant to join the union but didn't see any other way to change the situation. "I never liked those things because I heard when people unionized they would be killed," she told me. "I was so scared to be killed. They could kill my kids." (Her fears were perfectly warranted: from January 2015 to February 2018, at least thirty-one unionists had been murdered in Honduras.)

In fact, many of the striking workers received death threats and were physically attacked and intimidated, both by company security guards armed with high-caliber rifles and by agents of the national police. A report from the Network against Anti-Union Violence recounted eleven incidents of assault and/or intimidation among workers for the two companies. Mejía had been routinely harassed by the man responsible for supervising the guards. "He would say, 'If it was up to me, I would have gotten rid of you already.'"

Finally, in March of 2018, the workers negotiated the rehiring of most of the fired union members, but the company refused to reinstate seventeen of the most active ones. Grupo Jaremar also registered two yellow unions for recognition by the Honduran Labor Ministry, and, though it was required by law to engage in negotiations with STAS, the company refused to admit labor inspectors and repeatedly failed to show up at government-mandated mediations. Workers said they were pressured to affiliate with the company-controlled unions and were offered bribes to act as spies inside of STAS. "They gave me soft jobs, easier jobs," one reinstated worker told me. "They were trying to make me fall in love with them, so that I would go over to the yellow union." (One of the guys that had led us around the plantation had requested that we not post anything about the visit on social media. He hadn't spoken up during a union meeting the previous day, he explained, because he'd suspected that there were company spies in the room. He'd already been getting threats for having affiliated, including from one of the company's guards, an ex-military man who had made it a point to show him his gun.) The government of President Juan Orlando Hernández, whose tenure has been marked by violence and corruption, aligned with Jaremar in its effort to bust the unions, a move that in September of 2019 prompted fifty-five congressional representatives to call on the U.S. secretary of labor and the trade representative to hold the Honduran Labor Ministry accountable for having violated the Central American Free Trade Agreement, or CAFTA.

Just as Marquez's fictional town of Macondo is left in ruins after the banana company leaves, so have the palm oil communities in Honduras, and in neighboring Guatemala, found themselves worse off than before the industry's arrival. Farmers in both countries continue to be pushed off fertile land by plantations, forcing them to cut down yet more trees in order to grow food. The deforestation has in turn taken a toll on the local climate. Since 2014, a major drought affecting the region has been responsible for increasing numbers of Hondurans and Guatemalans migrating to the United States. In December of 2018, after a seven-year-old Guatemalan girl died while being held in detention by U.S. border agents in El Paso, Texas, her father said that the two had fled their village because deforestation to make way for oil-palm plantations had made subsistence farming there impossible. "The palm oil companies say they're bringing development," a forty-three-year-old Guatemalan named Maria Margarita Ivanez told me one afternoon in her village. "But the only ones seeing development are the *cantinas*, which at the end of the month are always full with people buying liquor, liquor, liquor."

Sitting in front of his cinder-block house in Campo Amapa, not far from the home of Walter Banegas, a thirty-three-year-old Grupo Jaremar veteran named Marcelino Flores told me that he had finally just given up on his former employer. Skinny in knee-length soccer shorts and a stretched-out T-shirt, Flores explained that he'd worked as a cutter for a decade, clocking in at 6 a.m. every Monday to Saturday, but had been making just $50 a week. When he joined the union, the father of three told me, he'd been harassed and threatened by the bosses. "You can work decades here," he said, "and it's just nothing. There's no way to save any money. It took me five years to build this house, with my nails dirty."

Finally, in November of 2018, Flores said goodbye to his family and caught a bus from San Pedro Sula to the border of Guatemala. From there, he boarded another bus and then hopped a train that took him all the way to Miguel Alemán, just across the Rio Grande from Laredo. He could see Texas on the other side, close enough for him to begin imagining the new life he would build there, one where he'd be able to take care of his wife and kids. But at nine o'clock on the morning he was meant to cross the river, he and the hundred or so other migrants bedding down in a shack were roused violently awake. A dozen Mexican soldiers knocked down the door and began pushing people around, guns poised. The women and children were screaming, Flores said, everyone scared out of their minds. Immigration officials marched the would-be migrants onto buses, and they were driven straight back to San Pedro Sula. Flores' brother in North Carolina had paid a "coyote" $7,000 to get him across the border.

Now, a month later, Flores said that he could think of at least ten other Grupo Jaremar workers who were planning to join one of the "caravans" then heading for the U.S. border. He himself was plotting another attempt in just a few weeks. "I couldn't finish building the floor of my house with the salary I get from the company," he said. "I don't want to know a damn thing about those palm trees."

7

The World Is Fat

People are fed by the food industry, which pays no attention to health,
and are healed by the health industry, which pays no attention to food.

—Wendell Berry, *Sex, Economy, Freedom & Community*

DR. ANOOP Misra drew back the flimsy curtain in his office, and the
patient stepped down from the exam table, gently tugging the bottom
of his shirt so as to obscure a considerable midsection. "I'm not here
to give you sweet words," said the soft-spoken endocrinologist, who,
in addition to seeing patients six days a week at this upscale health
center in New Delhi, chairs India's National Diabetes, Obesity, and
Cholesterol Foundation. Dressed in a white lab coat and with neatly
parted thick gray hair, Misra reclaimed his position behind the desk
and turned his attention to the patient's wife. What did she cook at
home, he wanted to know, and using which kinds of oils? "The diet
is all fried," the doctor told me after the couple had gone. "This man
is sixty-two and has already suffered a heart attack seven years ago."

Across the lobby of the bustling facility, where women in saris and men in sandals sat beneath signs reading "Advanced Centre for Insulin Pump" and "Centre for Metabolic and Weight-Loss Surgery," Shubhra Atrey, one of three clinical nutritionists who work with Misra, echoed her boss's dismay. South Asians are genetically inclined toward elevated risks of diabetes and cardiovascular disease, but in the seven-plus years she'd been practicing, Atrey said, she'd watched her fellow Indians undergo a transformation. "There's more obesity, including childhood obesity, and we have seen that obesity causes more diabetes." These days, she and her colleagues were seeing some sixty obese patients every day. "We counsel them. Ninety percent of the time, we discuss about oils. Bad oil, good oil. Palm oil is not a very good oil."

I had traveled to India to see for myself how the palm oil revolution was playing out in the country that imports more of the stuff than any other nation. A 2017 study in the *New England Journal of Medicine* found that the global prevalence of obesity and overweight had skyrocketed over the previous twenty-five years, with more than 10 percent of the world's population now considered obese. Some of the highest increases had occurred in developing countries, many of which were also confronting epidemics of under-nutrition. In India, non-communicable diseases such as diabetes recently overtook infectious ones such as diarrhea and tuberculosis to become the leading killers. Today, India has more patients with Type 2 diabetes than any other country in the world. (Yes, it also has more people than most, but those with diabetes form a disproportionately high percentage of its population.) The *Journal* researchers pointed to the "increased availability, accessibility, and affordability" of high-calorie foods to explain the worldwide packing on of pounds. "We have more processed food, more energy-dense food, more intense marketing of food products," Dr. Ashkan Afshin, the study's lead author, said upon its publication.

We also have more palm oil. During the years looked at by the researchers, 1980 to 2015, global production of the commodity increased more than twelve-fold, from 5 million to more than 62 million metric tons. The growth in production of palm oil has surpassed even that seen in wheat during the transformative period of the mid-twentieth century known as the Green Revolution. What happens to all that oil? Some 70 percent of it ends up in just the sorts of processed and "energy-dense" foods cited by Dr. Afshin.

"It is expected that 45 of every 100 additional calories in the period up to 2030 may come from oil crops," Carl Bek-Nielsen, the model-pretty Dane who now leads United Plantations, told an audience of industry executives in 2012. "Oil palm's contribution as a stabilizing crop to global food security is now undisputed."

Well, not exactly. While it's true that many of the world's people could use more calories—and certainly we all need some fat in our diets—the global glut of palm oil is in fact diminishing food security, in a fairly drastic way. It's common to blame sugar for the world's weight problems, but in the last half-century, refined vegetable oils have added far more calories to the global diet than has any other food group. Between 1961 and 2009, for example, the availability of palm oil worldwide went up a staggering 206 percent. Over the same period, the availability of sugar and sweeteners increased by just 20 percent. More recently, in the decades from 1991 to 2011, the global supply of food energy increased by 278 calories per person, with more than a quarter of that increase coming from vegetable oils. In South Asia, the oils accounted for 32 percent of the increase in consumed calories. But it isn't just the oils themselves. Part of the problem is the sort of nutrient-deficient, heavily processed junk that all of this cheap oil enables. And land planted with oil palm, of course, is land not being used to grow healthful foods such as fruits, vegetables, and legumes.

Over the last four decades, an influx of foreign investment into the food markets of poorer countries has helped fuel this trend. Between 1980 and 2000, investment from the United States into overseas food processing quadrupled, from $9 billion to $36 billion. Most of that cash came from multinational companies such as PepsiCo, Nestlé, and Yum! Brands, the Louisville-based owner of Pizza Hut, KFC, and Taco Bell—companies that make just the sort of ultra-processed foods for which palm oil is ideally suited. Just as farm policies in the United States led to the overproduction of corn and subsequent rivers of high-fructose corn syrup and endless conveyor belts of fast food in this country, so have international trade patterns abetted the palm oil bonanza, bequeathing a global landscape saturated in deep-fried snacks and fast and processed foods.

The implications for public health are enormous. Over the past decade, large-scale studies from France, Brazil, the United States, and Spain have echoed the *Lancet* in finding that the high consumption of ultra-processed foods is associated with higher rates of obesity. When eaten in large amounts, they have also been linked to depression, asthma, heart disease, and gastrointestinal disorders. In 2018, a study published in *The BMJ* (formerly *British Medical Journal*) found that a 10 percent increase in the proportion of ultra-processed foods in the diet led to an increase of more than 10 percent in the overall risk of developing cancer.

That said, the oil itself does raise some concerns. Part of what makes palm oil so useful to these brands is the fact that it is 50 percent saturated, which helps with both providing the desired "mouth feel" and prolonging the shelf life of products. As explained previously, palm oil's high smoke point makes it ideal for frying up chicken nuggets, French fries, cheese curls, and doughnuts—as well as such Indian staples as samosas and *poori*. (Palm kernel oil, at 80 percent saturated fat, tends to be used more by the makers of chocolates and other confectionery, who prize its hard texture, among other qualities.)

In 2018, following moves by the United States and other countries, the World Health Organization (WHO) announced an initiative aimed at eliminating artificial trans-fatty acids from the global food supply by 2023. The substance is produced when vegetable oils are partially hydrogenated, generally for use in processed foods. (Trans-fatty acids occur naturally in meat and dairy products from ruminant animals.) Nutrition experts applauded the effort to phase out the industrially produced fats, which have been linked directly to cardiovascular disease, but their common replacement—palm oil—presents problems of its own.

Public opinion has shifted somewhat when it comes to the relative dangers of saturated fats—replacing them with refined carbohydrates, for example, appears to do more harm than good—but studies have shown that diets rich in palm oil, which contains minimal amounts of omega-3 and omega-6 fatty acids, both of which have health benefits, lead to a higher risk of heart disease than those heavy in such unsaturated fats as olive or soybean oils. (When it comes to consuming palm oil in processed foods, eaters are in for a double whammy, as it tends to appear in concert with refined carbohydrates.)

In recognition of these findings, the American Heart Association released a Presidential Advisory on Dietary Fats and Cardiovascular Disease in 2017 in which it called for lower intakes of saturated fat and higher intakes of unsaturated fats. "Trans fats carry a bigger risk per gram than saturated fats," explained Barry Popkin, a professor of nutrition at the University of North Carolina–Chapel Hill, "but the volume of palm oil being consumed is so much greater." When it comes to the overall health effects, he said, palm oil may have a much more harmful impact.

This is the case particularly in the developing world, where palm oil continues to supplant other oils thanks to its low price. India, for instance, has seen palm oil, at $694 a metric ton, displace more traditionally used oils, including sunflower (currently $832 a ton), rapeseed ($890), and groundnut ($1,876), especially by the food industry, which, as we know, tends to buy a refined, bleached, and deodorized version of the oil. "Even in developing countries, excess calorie intake is a concern, let alone in terms of composition of that intake," said Dr. Qi Sun, an assistant professor of nutrition at the Harvard T.H. Chan School of Public Health. "In terms of palm oil, which is not healthy, I think the consequences could be devastating."

A ninety-minute drive from Delhi, the village of Taoru, in the Mewat district of Haryana state, is a jumble of two-story buildings in red brick and cement-block fronted by flashy commercial signage. On an apocalyptically smoggy day in October, the town was bustling, with vendors manning wooden carts piled with shiny orange persimmons and dwarf pomegranates, and honking motorbikes dodging donkeys pulling all manner of burden. Accompanied by a researcher from the Public Health Foundation of India, I wandered into one shop and inquired about purchasing some palm oil. The merchant said he didn't carry it. Only after my companion insisted that we needed something very cheap did the man fish out a red-and-yellow-striped packet of Ruchi Gold ("India's number 1 palm olein brand") from the recesses of his shelves. The local food vendors buy it, he told us, but it shouldn't be used at home, because it isn't good for health. In another

Street vendors in India use palm oil for frying snacks—though they don't always admit it.

shop, we asked what oils were available and were presented with a pair of liter bottles of mustard oil, priced at 90 and 115 rupees respectively ($1.21 and $1.54 at today's exchange rates). Wasn't there anything less pricey? The shopkeeper finally placed a plastic liter pouch of Ruchi on the counter, for 70 rupees ($0.94).

"People are using and selling a lot of palm oil here," a newspaper journalist named Adarsh Garg explained later, over cups of cardamom-scented tea. "Nobody tells, but they are using." It's little wonder: in a country where the average citizen earns less than $2,000 per year, it makes sense that people would seek out the least-expensive option; when it comes to a product many households will buy multiple times a week, a thirty- or sixty-five-cent price difference can add up quickly. India's street vendors—there are more than 10 million of them—also are increasingly reliant on palm oil. At a night market in the Chand Nagar neighborhood of New Delhi, amid pop-up stalls offering everything from lace bras and plastic shoes to eggplants and shallots, a twenty-four-year-old vendor named Ajit Yadav told me that he goes through five fifteen-kilogram tins of palm oil every week in order to churn out his popular *jalebi*, swirly-shaped, deep-fried sweets. "Of course the vendor will use it," Misra's colleague Amrita Ghosh told me. "He wants to earn money. He's not concerned about anyone's health."

It may be an open secret that India's street-food vendors fry their goods in palm oil, but the capital's wealthier denizens say they don't go near that stuff. Still, they're likely ingesting far more palm oil than they realize. Commercial *vanaspati*, the hydrogenated oil that has been swapped in for *ghee*, or clarified butter, over the years as dairy prices in the country have risen, increasingly gets made from palm. India's producers of blended cooking oils also are sourcing more of the commodity. "The cheaper oil is often mixed in higher proportion than the expensive oil," explained Pawan Agarwal, chief executive officer of the government-backed Food Safety and Standards Authority of India, or FSSAI, "while in the marketplace they suggest it is the other way around." Even those buying "pure" versions of such traditional oils as mustard, sunflower, and rapeseed—all of which are grown domestically (and which have saturated fat levels of 12, 10, and 7 percent, respectively)—will often go home with something adulterated with palm. "We cannot mention any brand in India that you can rely on," said nutritionist Atrey.

Kamal Kapoor wouldn't argue with any of this. For twenty years, the friendly owner of KP Agro Oils has been buying soy, mustard, and palm oils from multinational commodity traders like Cargill and Louis Dreyfus and selling them to manufacturers like Haldiram's, the India-based maker of the savory snacks known as *namkeen*, and to shopkeepers that supply street vendors. These days, he also sells to Domino's Pizza, McDonald's, and Carl's Jr., among other purveyors of fast food. Of the two thousand metric tons of oil he buys every year, said Kapoor, who welcomed me into his office, situated three flights up in the Tilak Nagar neighborhood of western New Delhi, nearly half of it is now palm oil. He opened a rusty metal file cabinet and pulled out a plate of nuts and a package of crumbly, palm oil–saturated snack mix. "There are a lot of new varieties of these snacks," he said, gesturing toward the latter. "New manufacturers coming."

Kapoor, who has deep-set eyes and dark hair that he feathers away from a center part, also sells palm kernel oil to sweets manufacturers and local bakeries for use in chocolate toppings, decorative icings, and non-dairy whiteners. These days, he said, it also gets used to make imitation dairy products. "A few years back, they made cake with fresh cream," he told me, "but now they make it with palm kernel oil. It's a secret," he added. "They don't write it on the label. How do I know? Because I supply them!" He let out a laugh.

Down a floor in the clanging packing area, shelves sat crammed with fifteen-kilogram tins of oils, mostly palm, which Kapoor sells for 1,050 rupees, as opposed to 1,200 for the soy oil and 1,250 for the mustard. He led me over to a foot-square box sitting on the floor and pulled back a cardboard flap, exposing a creamy mass encased in plastic lining. "Oscar vanaspati palm stearin" read the label. The hard but malleable block had the texture of children's modeling clay. "We don't prefer to sell this one," Kapoor said. "We have a bad conscience. Only D-grade people ask for this oil." The solid portion that results after fractionation is most commonly used for making soap, he explained. But in India, bakeries and local biscuit factories now use the *vanaspati* for cookies, cakes, and other sweets. "If you take one spoonful every day for a month," said Kapoor, "you're going to die. The companies know it, but they sell it only for money, money, money."

Indian consumers have learned to be leery of locally produced foods, but even the brands they tend to trust—multinational names like PepsiCo and Nestlé, McDonald's and Domino's—now deliver large quantities of palm oil into the national diet. Given India's 1.38 billion

people and emerging middle class, not to mention a young generation larger than the entire population of the United States, it's no
surprise that such companies are going big into the subcontinent.
In the five years between 2014 and 2019, India's "limited-service restaurants industry" (including the fast-food and takeout sectors) grew
by 86.3 percent, according to the market research firm Euromonitor. "Every five to ten kilometers, there's a KFC's or McDonald's,"
said nutritionist Atrey. Domino's Pizza now operates 1,328 stores in
India. Subway has 508. There are 430 Pizza Huts and 395 KFCs.
You'll find more than 300 McDonald's in the country.

At the same time, the vertical strips of perforated single-serve
snack packages in bright lime-green, fire-engine red, and lemon
yellow have become a fixture of the Indian landscape, whether in
clamoring cities or sleepy rural outposts. Between 2014 and 2019,
sales of packaged food in the country increased an eye-popping
125.1 percent, or 17.6 percent a year, according to Euromonitor.
Sanjay Kumar, the proprietor of Sanjay General Store, a ten-foot-
wide establishment in Taoru, told me that he now sells fifty to sixty
packets of processed snacks every day—fully half his total sales. A
decade ago, such purchases made up only a tenth of his business.
The customers are mostly kids, Kumar said, who are attracted by
the colorful packaging and who've seen the products advertised on
TV. Priced at no more than 10 rupees (15 cents) a piece, the snacks
invariably feature "palm olein" as their second-most-prominent
ingredient. Lay's Classic Salted Chips, Kurkure Masala Munch,
Uncle Chipps Spicy Treat (all PepsiCo brands); Haldiram's SnacLite
Fries and Navaratnam; Bikano brand Natkhat Nimbu Lemon Hit:
every one of them lists the substance among their first two ingredients. Not just chips, but cookies, too. Poof Strawberry Wafletts?
Third ingredient. Karachi Bakery Almond Cake Rusk? Third, after

flour and sugar. The single-serving packets of Wai X-press noodles, priced at 10 rupees? "Edible vegetable oil (palm)": second ingredient. India's wildly popular Maggi instant noodles, which are made by Nestlé, also list palm oil second. The place is drowning in the stuff.

Just as palm oil companies operating in countries with weak governance can get away with labor and environmental abuses, so too are multinational food corporations growing their businesses in places where health infrastructure is less established and the population less informed when it comes to nutrition. Indeed, like the tobacco and soda industries before them, the world's junk- and fast-food purveyors appear to be pushing these products on the developing world while fully aware of the risks they pose.

PepsiCo India, a main player in the country's snacks sector, had in fact launched a health-focused "Snack Smart" line of Lay's, Kurkure, and Cheetos snacks back in 2007, along with a high-profile campaign touting how it had slashed saturated fats in the products by replacing palm oil with rice bran oil. But in 2012 the company quietly removed the Snack Smart logos, having reverted to palm oil as a way of cutting costs. The move followed a similar, brief shift by the company to a healthier blend of oils in the Cheetos it sells under the Mexican brand Sabritas. "Higher commodity costs required a return to palmolein," wrote PepsiCo senior vice president and chief scientific officer Mehmood Khan in a 2010 internal review.

"Oil is an expensive part of junk food," said the University of North Carolina's Popkin. "It's not like the sugar in Coke." The multinationals are cost-conscious, he continued, "but it's more important for something that costs four, five, ten percent of the food than it is for something that costs a tenth of a percent."

On the PepsiCo website, readers can learn that the 28.3-gram bags

Palm oil–laden snacks are readily available even in India's remote rural outposts.

of Crunchy Cheetos it sells in the United States have 1.5 grams of saturated fat, or 5.3 percent, thanks to being cooked in sunflower, corn, and canola oils. The Cheesy Krunchy Cheetos it sells in India, on the other hand, have a saturated-fat content of 5.31 grams, or "not more than 17.7% by weight," in a single serving. That's about a fifth of the total daily intake recommended by the WHO, whose guidelines the company purports to follow.

PepsiCo is well aware of the risks. "The scientific evidence linking dietary saturated fat intake to atherogenesis [plaque buildup in the arteries] is strong and compels the transition from animal fats and tropical oils to healthier oils," wrote Khan in that 2010 memo. "Among the many actions the food industry can take," he continued, "the most important include an effective transition from saturated tropical oils to healthier oils." But a decade later, PepsiCo continues to help keep costs low by sourcing massive volumes of palm oil—485,756 metric tons in 2019. In September of 2019, PepsiCo India announced plans to double its snacks business in India by 2022.

In 2011, Yum! Brands announced that it was removing the palm oil from its deep-fat fryers in its U.K. KFC stores in order to achieve the "double benefit" of reducing heart disease and minimizing its contribution to climate change. The following year, KFC Australia changed its cooking oil from palm to "locally sourced high-oleic canola oil," and in 2015, the company wrote on its website that its "goal over the next two years is to phase out palm oil wherever feasible." But in 2019, Yum! used 180,000 metric tons of the oil. Its updated website no longer says anything about health concerns related to the ingredient. When asked to provide specifics on the countries in which it uses palm oil, a spokesperson wrote in an email: "Yum! Brands is committed to sourcing more sustainable ingredients for the foods we serve to our consumers globally. We have reduced the amount of palm oil used globally in our cooking oil and continue to be focused on sourcing certified sustainable palm oil for the remaining palm oil we use for cooking."

Meanwhile, unlike those in the United States, Yum! customers in most of the roughly 140 countries and territories in which it operates will not find nutrition information in its stores. In some cases, they can go online to obtain calorie counts, but nowhere will they find product ingredients or levels of saturated fat. The spokesperson declined to provide this information.

Like its fast-food peers, McDonald's, which purchased 92,534 metric tons of palm oil in 2019, relies on the commodity in some regions more than in others. The company currently uses palm oil for frying "in many of its Asian markets," according to a spokesperson, and to par-fry chicken and potato products by direct suppliers "in certain markets." Ingredient listings for overseas restaurants are available neither in stores nor online. (The spokesperson suggested that customers request this information from a restaurant manager or via the "contact" section of the website.)

It's more of the same in Latin America. After the signing of the North American Free Trade Agreement, or NAFTA, in 1994, a flood of direct investment flowed into the region's food-processing industry, with sales of junk food in Mexico, for instance, growing by 5 to 10 percent annually between 1995 and 2003. Today, the country is one of the ten biggest producers of processed foods in the world. The companies behind its popular junk foods, including such usual suspects as Nestlé, Unilever, PepsiCo, and Yum!, as well as Mexico's own Grupo Bimbo, are among the world's top purchasers of palm oil. Since 1995, palm oil imports to Mexico have increased more than five-fold, from 103,000 metric tons to 565,000.

Whereas in 1980 only 7 percent of Mexicans were obese, by 2016 that figure had tripled, to 20.3 percent. As in India, diabetes is now a top killer in the country, claiming some eighty thousand lives every year. Today Mexico has some of the highest rates of obesity and high blood pressure in the world, with 29 percent of its children between the ages of five and eleven overweight, as well as 35 percent of kids between eleven and nineteen. In a 2018 study looking at changing diets in Latin America and the Caribbean, UNC's Popkin found that since the 1980s there had actually been a decline in carbohydrate intake but noticeable increases in total fats in the diet. The largest increase? Vegetable fats.

"This is an equity issue," said Saskia Heijnen, who oversees the Our Planet, Our Health program at the London-based Wellcome Trust, which has funded research into the environmental and health implications of the global palm oil industry. "The people who can afford it can buy products with less palm oil or no palm oil, whereas the people who can't are stuck with it."

Nor is palm oil doing anything close to stabilizing food security at its source. In Sumatra, I spent time with indigenous Orang Rimba and Batin Sembilan communities who said that the diets they had relied on for generations were simply no longer available. Staples including cassava, sago (the palm starch that had originally lured Fauconnier east), wild boar, deer, squirrel, snails, mushrooms, ferns, fruits, and caterpillars—wild foods rich in vitamins and micronutrients like iron and calcium—had disappeared when the forests fell. These days, the communities rely on instant noodles (containing some 20 percent palm oil) and government-issued rice to survive. The single shop that operates near a settlement I visited inside a concession in Jambi province rarely had fruits or vegetables. Instead, brightly colored packages of candy and fried snacks shared the shelves with multiple brands of clove cigarettes.

A 2019 study by the Indonesia-based Center for International Forestry Research, or CIFOR, found that in that country, diets in communities where the palm oil industry had moved in were far less healthy than those in traditional communities living in the same region. In Borneo's West Kalimantan, for instance, indigenous Dayak people practicing traditional agriculture consumed more fruit and fish than people living in villages where oil palm was grown. Children from traditional households ate substantially more fruits, vegetables, fish, and staple foods than those living in oil-palm communities. The rate of wasting (the weakening of the body due to malnutrition) among children under five in oil-palm communities was higher than that in traditional households. At a similar site in Papua,

the researchers found a higher rate of anemia among mothers working in oil palm than among those in traditional households.

On the outskirts of a national park in Jambi province, I spoke
with Hassan Basri, the sixty-two-year-old chief of an Orang Rimba
clan now scraping by in a clearing of sparse trees amid structures of
sticks and plastic tarps. "We have been here for many, many generations," he told me. "We are not newcomers." Still, his people had lost
everything about their traditional livelihoods to the palm oil industry. Even four years ago, said Basri's visibly angry stepson, the men
in their clan had hunted wild boar here. "Now the forest is gone,"
lamented Basri, a bony Giacometti come to life. "We can no longer
provide for the community."

In the past, the province's indigenous populations would collect
rattan and "dragon's blood," a bright-red resin used to make dyes and
incense, which they would trade with neighboring villages for food
and other staples. But they can't source these materials anymore, nor
can the women pass on intergenerational knowledge and skills, such
as weaving and cooking forest fruits, vegetables, and roots, to the
younger generation. Driving through a government-owned oil-palm
concession in northern Jambi, I passed a pair of Orang Rimba wom-
en, babies tethered to their backs, begging with tin cups by the side
of the road.

"My children are like refugees compared to when I was young and
could walk in the woods any time," Abdullah Sani, a fifty-three-year-
old Batin Sembilan father of four, told me. "Rambutan. Jackfruit.
Durian. We used to get all of them. We never had to buy rice or
vegetables. We used to fish in the river. We would get wood from the
forest for our houses. Now we have to buy it. We used to get river
catfish. Now the river is polluted. The quality of the water has been

changed because of the chemicals. Now we have to buy water. Do you think this water can be consumed?"

Locals said that only one hardy species of fish managed to survive in the tea-colored river, thick with runoff from the plantation and contaminated, they said, with fertilizer and pesticides. Residents of Sayaxché, a municipality in Guatemala's Petén region, told me the same thing. Since the effluent pools of the local palm oil company had overflowed in 2015, they explained, one type of fish had displaced all the others that once thrived there. They sent a teenager off to the river to fetch the fish they call "*el avión*" as proof. Cylindrical and solid, like an armored military aircraft, the ungainly creature had whiskers and a mottled beige-and-black skin that was rough to the touch. The "airplanes," which grow as long as a foot, bite at the women's feet when they go to do the wash, and their sharp fins leave bloody cuts on their ankles. "These are the only fish we can catch, these devil fish," said a local named Remigio Caal. "You can't eat them. They're all bones."

Back in India, Food Safety and Standards Authority CEO Agarwal told me that his government was taking steps to address his country's obesity problem, including through a focus on processed and fast foods and, in particular, on unhealthy ingredients. Already, said Agarwal, who met me in his sprawling New Delhi office, his agency had set a two-percent limit on trans fats in *vanaspati* and bakery shortenings. In 2018, it issued a draft of rules requiring manufacturers to display the fat, sugar, and salt contents of products on the front of packaging—including a red flag on any that exceed certain thresholds for salt, sugar, and saturated fats. But later that year, after food companies expressed concern about the proposals, the FFSAI delayed implementing the system, instead establishing a three-person

The fish the locals call "*el avión*"—the only one hardy enough to thrive in the polluted river.

panel to review the rules. Among those appointed to the panel was Dr. Boindala Sesikeran, a former adviser to Nestlé and to Nutella-maker Ferrero, and a trustee of something called the International Life Sciences Institute, an American nonprofit known for advancing the interests of the four hundred corporate members—among them PepsiCo, Nestlé, ADM, and Cargill—that provide its $17 million budget.

The industry has likewise lobbied against the possibility of a national junk food tax. In 2013, Sanjay Basu, an assistant professor of medicine at Stanford University, conducted a study that determined a 20 percent tax on palm oil in India would avert some 363,000 deaths from heart disease over ten years. (In 2015, the Singaporean government introduced a "healthier ingredient scheme" in which it began subsidizing healthier alternatives to palm oil for street vendors in that country.) Given the size of its informal foods sector, enforcing such a tax in India would pose a formidable challenge. In a country still facing widespread malnutrition, raising prices on any food source also introduces ethical questions.

"There are some nutrients that you're getting from palm oil," explained Shauna Downs, an assistant professor at Rutgers School of Public Health. "Like, you are getting fat, which you need." Downs wrote her PhD thesis about policies that the Indian government might adopt to address the rise in non-communicable diseases in the country. "Still," she said, "there's a difference between food security and nutrition security. Not every calorie is created equal. People need calories, but do they need to have a huge proportion coming from palm oil? No."

Whatever opposition local players and multinational companies mount on the ground in India, Mexico, and elsewhere, it will undoubtedly have the backing of the palm oil lobby, which, after all, has the responsibility of finding a market for its ever-growing volumes of oil.

After my time in New Delhi, I caught a flight to Jakarta, where I joined some seven hundred participants at a forum aimed at revolutionizing a global food system that, according to its keynote speaker, a Norwegian physician and environmentalist named Gunhild Stordalen, "is failing both us and the planet." Among the presenters at the two-day affair was the minister of health for Malaysia, Dr. Subramaniam Sathasivam, who gave a speech in which he lamented that unhealthy diets were leading to increasing numbers of deaths globally and that agricultural monocultures were adversely impacting the environment. It would have been good, he said, had the Malaysian and Indonesian ministers for agriculture and finance also been in attendance, given the roles they could play in fixing such a broken system.

I emailed Sathasivam's press secretary afterward to arrange for an interview and received an eager reply—until I mentioned in passing my interest in his country's main export crop. "He can't talk about palm oil," I was told. "He can talk about anything but that."

8

Smog Over Singapore

I heard the sound of a thunder, it roared out a warnin'.
Heard the roar of a wave that could drown the whole world.

—Bob Dylan, "A Hard Rain's A-Gonna Fall"

THE BUREAUCRAT sat behind his desk, smoking. I'd been summoned to the sour-faced man's office, a ridiculously large, and largely unfurnished, room on the second floor of an administrative building in the Sumatran city of Jambi, to explain my presence in his district. Accompanied by a few staff from a local NGO, and with binoculars none-too-subtly hanging from my neck, I handed him photocopies of my passport and headshot and went over my lies in my head—*here to see hornbills and orangutans, here to see hornbills and orangutans*—while trying to maintain my composure.

Three weeks before my departure from New York, an article about a journalist dying in an Indonesian prison had turned up in my Twitter feed. In the four months before his death, Muhammad Yusuf had written more than twenty stories about a controversial oil-palm plantation and the powerful tycoon behind it. The forty-two-year-old had been held for several weeks on charges of defaming the company, and though the official cause of Yusuf's death was heart attack, his wife had been denied access to the body. Handwritten autopsy notes obtained by a local news outlet cited extensive bruising on the journalist's neck, shoulders, back, and thighs, and the national human rights commission had vowed to investigate.

Of course, I'd planned this trip, as I had my previous ones to the region, with the express purpose of reporting about palm oil. The industry, as we know, is a colossus here in Indonesia, where the tolls it's taken on indigenous communities and wildlife are cause enough for alarm. But now I wanted to investigate its broader impact on the planet. The links between tropical deforestation and carbon emissions are well established. So why, I wanted to find out, was it proving so difficult to stop the destruction? I knew that not everyone would be pleased with my line of inquiry. In 2014, the American actor Harrison Ford had taped a confrontational interview with Indonesia's environmental minister for the Showtime series *Years of Living Dangerously* and was subsequently threatened with expulsion from the country. Two years later, after Leonardo DiCaprio exhorted his then–15.8 million Twitter followers to sign an online petition demanding that President Joko Widodo protect Sumatra's biodiverse Leuser Ecosystem, government officials had accused him of running a "black campaign" against the palm oil industry and threatened to toss him out, too. Still, this dead journalist suggested a new level of menace.

In the early 2000s, the Indonesian government passed a number of reforms aimed at replacing its previous system of centralized power—never a great fit for a country of some seventeen thousand islands—and devolving political authority to the provincial and

county levels. Since then, the ability to award logging and plantation licenses has rested with county governors, known as *bupatis*. A
few years later, the convergence of high global prices for fossil fuels
and sagging ones for agricultural commodities had led governments
around the world to begin focusing on biofuels. George W. Bush,
citing America's "addiction" to foreign oil, vowed in his 2007 State
of the Union address to cut gasoline usage in the United States by
20 percent within a decade. "We make a major step toward reducing
our dependence on oil, confronting global climate change, expanding
production of renewable fuels, and giving future generations a nation
that is stronger, cleaner, and more secure," he said later that year,
upon signing the Energy Independence and Security Act. The bill,
which passed with bipartisan support—House speaker Nancy Pelosi
called it "groundbreaking"—set higher fuel-economy standards for
cars and light trucks and required the production of 36 billion gallons of renewable fuels (incorporating corn, soy, oil palm, and sugar)

An oil-palm plantation on the island of Sumatra.

by 2022, nearly a fivefold increase. Two years later, Europe followed with its own biofuels mandate, the Renewable Energy Directive, or RED, aiming for 10 percent of transport fuels to be derived from biofuels by 2020.

The palm oil industry in Southeast Asia received news of the mandates with glee. Having long lobbied Western politicians to introduce biofuels measures, plantation companies immediately set about expanding their production capacity even more. Indonesia announced that it would convert an additional 13 million acres of forest—nearly the size of West Virginia—to oil-palm cultivation. By 2011, imports of the commodity to the European Union had soared by 15 percent, followed by 19 percent the year after that. (Today the EU is the second-largest importer of palm oil, after India.) By 2017, oil palm was accounting for 31 percent of biofuels feedstock worldwide.

Securing the permits necessary for all that expansion now meant cozying up to the newly empowered *bupatis*. Bribes became central to the process, with politicians green-lighting projects involving friends, relatives, and anyone else willing to pony up the cash. (That dead journalist? He had been investigating a company owned by the nephew of a local *bupati*.) Often the arrangements involved skipping right past the required environmental and social-impact assessments, including the mandated consultations with communities likely to be impacted by the deals. When the consultations did take place, they tended to involve deception, with villagers signing over land rights in exchange for farming plots that never materialized and promised payments that never arrived.

The rampant illegality eventually led to the founding of something called the Corruption Eradication Commission, or KPK, the existence of which pushed the *bupatis* to engage in still more elaborate schemes. It became common for those in power to issue licenses to shell companies established by their acquaintances. Those friends would then turn around and sell the "shadow companies" to plantation firms, generally for hundreds of thousands of dollars, with much of that cash ultimately flowing back to the *bupatis*. One governor on Borneo set up no fewer than eighteen shell companies in the names of his relatives and cronies and then granted all of them licenses to establish large plantations. Those companies were quickly sold to the likes of Wilmar and other established firms. "All we got was oppression," said a farmer named James Watt, whose acreage was among that taken in the deals. "Clearing our land, dumping waste in our rivers. . . . It was always empty promises with [the *bupati*]. I think he saw being *bupati* as his chance to make as much money as possible."

I had heard about a Sumatra-based watchdog organization that tracks illegalities in the palm oil industry, and I figured that if anyone had a handle on the hidden forces at play, they would be the ones. The investigators for Eyes on the Forest, or EoF, spend weeks at a time in the remote regions of Indonesia where so much of the deforestation takes place. Deploying drones, satellite imagery, and finely honed undercover skills, they document how illicit oil-palm fruit makes its way to local mills and refineries and ultimately into our own kitchens, bathrooms, and fuel tanks. They'd agreed to show me how it was done.

"Make a right up here," said Wawan, gesturing out the windshield. Our driver, a nicotine-fueled kid in a backward-facing baseball cap and Vans sunglasses, eased the Toyota onto a rutted road, and we bumped along for a few miles before stopping alongside an electric-blue house set up high on stilts. Driving out from Jambi—it had taken more than a day, but I'd eventually gotten clearance from the bureaucrat—I had watched as open-fronted shops and gold-domed mosques gave way to oil palms extending in all directions. The guy I'm calling Wawan (all of the names here are aliases), EoF's lead investigator, had led us to this spot to see one of the farmers whose operations he'd been monitoring. The so-called encroacher had cut down forest in a protected area and was now cultivating oil palm in clear violation of the law.

Two men rose from the wooden platform-cum-bench on which they'd been sprawled and greeted us with handshakes and smiles. We pulled up plastic chairs, and they reclaimed their bench, each reaching for one of the Indonesian body extensions that is the clove cigarette. (The country, as mentioned, is the smoking capital of the world, with more than half of males over the age of ten lighting up every day. This despite the nausea-inducing images of ravaged throats and lungs that grace every package.) Shifting effortlessly into the farmers' Javanese dialect, Wawan, who speaks five languages, chatted with the pair about the weather and the elephants that had recently trampled through here, decimating the guys' young plants.

As we drove away an hour or so later, Buyung, an EoF staff member serving as my translator, explained that the men at the house believed Wawan to be a conservation expert tied to one of the local palm oil companies. He'll drop by from time to time, ostensibly to offer tips on maintaining the soil or dealing with pests, in order to keep up on the people cutting and planting in the area—and on the people behind them bankrolling the destruction.

The convergence of biofuels fever in the West and a newly decentralized Indonesia would eventually yield the proverbial perfect storm. Between 2000 and 2012, the country lost more than 15 million acres of natural forest, with its annual deforestation rate surpassing that of Brazil for the first time. Some three-quarters of Indonesia's current palm oil production has come online since 2000, with most of the growth happening here on Sumatra and on once-heavily-forested Borneo. Plantation firms now control more than eighty thousand square miles of Indonesian land—12 percent of the country. (In 2006, Indonesia once again overtook Malaysia as the world's number-one producer of palm oil, a title that it's held ever since.) "You once had just Suharto and his cronies stealing the country's natural resources," Glenn Hurowitz, chief executive officer of the Washington, DC–based non-governmental organization (NGO) Mighty Earth, told me. "There's only so much that one person can steal. Now you have five hundred little Suhartos stealing Indonesia's natural resources."

Over the years, the corruption has become increasingly intertwined with the country's elections. These days, running for even the most local office requires putting up hundreds of thousands, if not millions, of dollars, first to secure a place on the ballot and then to underwrite the campaign. Would-be candidates strike deals with the region's entrenched political parties—known locally as the "land mafia" for their intimate connections to the palm oil and other extractive industries—who help them get elected in exchange for lucrative contracts, jobs, and other benefits on the other side. Working through middlemen, they routinely buy the support of village chiefs, religious leaders, and other local powerbrokers and underwrite rallies, concerts, and other events, where they often hand out free meals.

When, in 2017, Abdon Nababan, the head of the country's main advocacy group for indigenous rights, attempted to run for governor in his home province of North Sumatra, he was informed that he would need either to pay a party millions of dollars to back his candidacy or, if he chose to stand as an independent, to collect the signatures—and photocopies of the ID cards—of 800,000 voters.

Through an intermediary, a consortium of business interests offered to provide 300 billion rupiah ($21 million) to bankroll Nababan's campaign, on the agreement that, once elected, he cede control of all land allocations. Other middlemen told Nababan that they would provide the required signatures—in exchange for 40 billion rupiah, or $2.8 million. He eventually withdrew from the race, leaving the field to a retired army general and a local palm oil baron. "I tried to get information about . . . who was behind this offer," he told a local reporter in regard to the signatures. "I came to the preliminary conclusion that they come from the oil palm sector, mining, and also the property business."

The corruption often goes straight to the top. In 2013, the chief justice of Indonesia's highest court was arrested after being caught taking a $250,000 bribe to decide an election dispute in favor of an incumbent district chief from Borneo. Local reporters traced the money back to a series of land deals, cut in the months leading up to the election, that placed a huge swathe of land inhabited by thousands of indigenous people under the control of a Malaysian-owned palm oil company. The judge and two associates were convicted in 2014, but the land deals were never revoked. A few years later, a Kalimantan-based *bupati* was caught granting oil-palm concessions to family members, who then flipped them to the companies that had paid for his election. Between 2014 and 2019, the KPK brought corruption charges against 240 sitting members of parliament.

The country's genocidal crimes having gone unpunished (and largely unacknowledged), a culture of impunity adheres, one in which intimidation still reigns. That land mafia, which is based in Medan, is composed of a handful of powerful families closely associated with paramilitary organizations, including the Pemuda Pancasila, the members of which are featured in Joshua Oppenheimer's

The Act of Killing, recounting the murders they'd committed during the 1965 atrocity. Renowned for its ruthlessness, the group has branches in villages across North Sumatra and functions as a sort of gang, operating under the guise of civil service. Driving around the region today, you still see their aboriginal art–like markings, painted on wooden posts along the road. "A lot of what they do is 'security,'" a New Zealander expat explained. "From themselves. As in, 'Nothing's happened this month; pay us, or something will happen next month.'"

While on an earlier reporting trip to Sumatra, I'd gotten stranded one night deep inside an oil-palm plantation with an Indonesian photographer as the sun was going down, and we ended up spending the night on the floor of a local chief's house. At one point in the evening, we heard voices and looked outside to see that a uniformed member of the Mobile Brigade Corps, a unit of the national police, had settled in on the patio, gun perched at his side. It was only after observing how the photographer, a burly guy in his forties, responded to the situation that I understood just how tense life there actually was. His palpable fear—Would they confiscate his camera? Throw us both in jail? Something far worse?—suggested a level of terror I'd only ever witnessed in Robert Mugabe's Zimbabwe.

If the industry flouted environmental laws before, the biofuels craze made things even worse. With demand for palm oil rising, everything, including the country's carbon-rich peatlands, had become fair game for development. As explained by Ian Singleton, this dense, waterlogged terrain comprises layer upon layer of dead organic matter built up over thousands of years. To ready it for planting, the palm oil companies dig deep trenches and then bring in machinery to fell the trees. Fires set to clear any remaining vegetation can continue to smolder for years. The Intergovernmental Panel on Cli-

Burning peatland on Sumatra.

mate Change (IPCC) has estimated that the carbon dioxide released
from peatlands—which in Sumatra and Borneo can run as deep as
sixty feet—amounts to some two and a half times that released from
cutting down an average (already-carbon-dense) tropical rainforest. "Once you start talking about saving Indonesia's peat swamps,"
Singleton told me, "it's not just about saving the orangutans or this
frog or that plant species. This is a global issue. You destroy all of
Indonesia's peat swamps, Planet Earth becomes uninhabitable. It's
that serious. It's not fucking around."

And yet destroying Indonesia's peat swamps is exactly what was
now happening, to the growing dismay of the international community. In 2011, partly as a response to the outcry over its ballooning
carbon emissions, the Indonesian president announced a moratorium
on new permits for clearing primary forests and peatlands for plantations and logging. The process involved a massive surveying effort,
under which millions of acres were designated as protected peatland.

But the mapping, too, eventually fell prey to the shenanigans of politics and the industry. A six-month review process put in place enabled *bupatis* to rezone areas from "forest estate" to "non-forest estate," thereby qualifying them for previously off-limits development. New maps routinely materialized in which borders had been inexplicably redrawn so as to redefine peatland as forest suited to "other use," including clearing for oil-palm plantations. One scam resulted in the 2014 arrest and conviction of a former governor of Riau province, who was found guilty of taking 1.5 billion rupiah (about $100,000) from a palm oil businessman in exchange for rezoning swathes of forest to qualify for oil-palm development. A 2018 Greenpeace analysis found that in the years since the moratorium went into effect, no fewer than 17,400 square miles of forest and peatlands had been removed from its original maps.

By 2013, the industry also had been forced to address its unsustainable ways. Wilmar International led the way with a pledge to avoid deforestation, planting on peatland, and exploitation in its palm oil operations. Other companies involved in producing, trading, and buying the commodity followed with their own such pledges, known as NDPEs for No Deforestation, No Peat, No Exploitation.

Nevertheless, the summer and fall of 2015 saw fires traced to oil-palm plantations across Sumatra and Borneo burning more than 6 million acres of forest, an area larger than the state of Vermont. Blanketing an expanse of Southeast Asia in haze for weeks, they sickened hundreds of thousands of people. A study published in *Scientific Reports* found that the carbon emissions released by the fires in September and October of that year were higher than those of the entire European Union over the same period. More than one quarter of the area that burned was located on land once included in the moratorium.

"Hey, sorry," wrote a Sumatra-based translator I'd worked with in an October email to me. "I was away longer than intended, trying to fly back to Jambi but the plane can't land cause of the bad haze. Some rubber, acacia, and palm companies are burning the forest (cheaper than cutting) to clear the land for plantation, taking advantage of the dry season. One of the companies is burning 62,000 hectares as far as I'm informed, and there's still dozens of other companies doing the same thing. Then they just paid off the corrupt cops here and it's all good." A few days later, he emailed again to say that he'd been hospitalized with a respiratory infection. A KPK audit carried out in response to the fires found that most had been set to clear land for planting oil palm, much of it on peatlands.

In the video, a man wearing jeans and a white tank top emerges from the driver's side of an empty dump truck and slyly passes a wad of bills to a hard-hatted guard before climbing back behind the wheel and driving off. Wawan had recorded the footage from a vehicle parked outside the entrance of a palm oil mill, where he'd arrived after trailing the truck, originally by motorbike, from a plantation located on protected land. (We would do a 4x4 switcheroo ourselves at one point, as a way of ensuring that a truck driver wasn't onto us. When the guys are traveling by motorbike, they'll often swap out shirts for new ones in different colors.)

As proof of their sustainability bona fides, companies with NDPE policies will often assert that their palm oil supply is "traceable to the mill." The idea is that, given the perishability of the fruit, which we know begins to degrade within forty-eight hours, whatever arrives at a facility must have come from within a certain geographical radius. The mills vouch that the nearby plantations from which they source are involved in neither the destruction of primary rainforest nor the draining of peatland.

But the model is ripe for abuse. As EoF reports have documented, drivers carrying fruit from illegal plantations will routinely race through the night to reach mills outside of their expected ranges. Sometimes they'll change license plates along the way. And we saw in Wawan's video what happens when they get there. "The cash is provided by the plantation owner," he told me. "It's part of the service. You give money to security to ensure that no questions are asked."

In order for Indonesia's now-very-significant milling and refining infrastructure to operate efficiently, it requires a continuous supply of raw materials, a situation that encourages buyers to look the other way where fruit provenance is concerned. Though it's generally assumed that plantation companies are the ones responsible for clearing the forests, as the population on Sumatra has increased and arable land become more scarce, individual farmers also are venturing farther into marginal areas, including peatlands. If those farmers set fires to clear the land for oil palm, they do so in the knowledge that whatever fruit they eventually harvest will find a ready market—regardless of how or where it's been cultivated.

Just to reiterate, growing oil palm isn't like growing basil. Significant cash is required up front to pay for things like fertilizers and pesticides, and the fruit isn't ready to harvest for at least three years. Cash-poor farmers looking to cultivate the crop often are compelled to sign on with landowners, generally absentee, who may promise to bequeath them small parcels of land in exchange for years of working a larger plot. Often the laborers—including those two guys on the platform—have no idea that they're cultivating on illegal land. Though the industry makes repeated reference to the fact that "40 percent" of Indonesia's oil palm is cultivated by smallholder farmers, the KPK has found that the actual figure is closer to 25 percent. And even those tend to be closely tied to big industry players.

In the years of Transmigration, the newcomers here in central Sumatra came mostly from Java, but these days, the arrivals to Jambi and Riau provinces hail mostly from up north, where, thanks largely to the industry, little land remains for families hoping to eke out a living. One afternoon, while driving across a remote landscape of oil palm and more oil palm, we watched as a giant bus, like something you'd see barreling toward Atlantic City on the New Jersey Turnpike, came flying around a corner kicking up dust. "Encroachers from North Sumatra," said Buyung, as though the presence of a crowded commercial vehicle here in the absolute middle of nowhere were a perfectly normal occurrence.

Given the shell companies, backroom deals, and looming threat of violence, tracking Indonesia's land transactions is no simple enterprise. Wawan and his colleague Wari will generally spend two or three weeks at a time on the road, often traveling by motorbike, and will pose variously as fishermen, birdwatchers, students, or land-scouting businessmen depending on the situation. Instances like the one with the guys on the platform, where they're hanging out with the very folks they're investigating, are not uncommon, and in extremely remote areas they may end up spending the night on such suspected criminals' floors. "If you go there openly," said Buyung of the serial deception, "they will bring you to the good thing. They'll say, 'Oh, we don't do that bad thing.'"

The job involves some next-level improvisation. Wawan will routinely instruct the driver to pull over so that he can chat up some trucker idling by the road or loading oil-palm bunches onto his rig. At one point, positioned in the middle of a plantation so as to monitor the trucks rumbling past, Wari had propped open our hood. Should anybody ask, we had broken down and were waiting for a

friend to arrive with a spare part. Some years back, while eavesdrop-
ping in a rural lunch spot, Wawan had heard people referencing the
businessmen behind the local deforestation and had recorded their
names in the squares of a prop crossword puzzle. Another time, having positioned himself near the entrance to a sawmill—before the
scourge of palm oil there was the wholesale heist of hardwood—he
tracked the number of flatbeds clearing the gates by transferring
matchsticks from one jacket pocket to another. The matchbox, too,
was a prop: the fifty-year-old father of two does not smoke. With
his trim blue jeans and thick-frame glasses, in fact, Wawan is like
EoF's resident dad-in-chief, a consummate mensch—aside from the
routine dissembling.

During our weeklong trip, the guys spent most mornings bent
over their laptops, studying truck routes and comparing satellite
images captured over time. In addition to being badass spies, they are
unapologetic tech nerds, conversant in the likes of GIS and eCog-
nition and engaged in, among other international collaborations, a
years-long mapping project with Google. (EoF, which was established in 2005, grew out of the Forest Crimes Unit of the WWF, or
World Wildlife Fund, with which it remains closely affiliated.) One
of the investigators, a physics grad named Jojo, builds his own drones
and travels with a 3D printer for fabricating broken bits on the fly.
(I'd asked Buyung what was up with the bizarrely long pinky nails
I'd noticed on so many Indonesian men, and he explained that they
were useful for fixing things, from crashed drones to crapped-out
motorbike engines.) The deskwork both informs the investigations
and adds a layer of evidence. "If somebody says, 'But I planted this
oil palm seven years ago,'" Buyung explained, "we can look at the
images and say, 'You lie.'"

They have been repeatedly harassed and threatened—their fre-

Eyes on the Forest investigators undercover in Sumatra.

quent covers as consultants for WWF often are enough to inspire ire. One former investigator had been forced to relocate his whole family. In 2007, Wawan was kidnapped and beaten by an angry mob egged on by a forest-service employee whose illegal business doings had been exposed by an EoF report. His kids broke down in tears when he showed up at home covered in cuts and bruises. It was only in the last few months, after his son had turned eighteen, that Wawan had come clean with him about the true nature of his work.

In recent years, it's become clear to legislators in the United States, as in Europe, that their biofuels initiatives were perhaps not the win-win-win scenarios that they'd initially imagined them to be. The idea behind "green" fuels, of course, is that by using plants to create energy, you expend only as much carbon burning them as was absorbed from the atmosphere during their growth. But because Americans weren't going to suddenly begin eating less processed food, the corn and soy oils that had previously gone into those products but were now being diverted to fuels needed to be replaced. (In 2012, the U.S. Environmental Protection Agency had determined that palm oil did not qualify as a sustainable fuel stock under the guidelines of the Renewable Fuel Standard.) That meant finding more land on which to grow those crops or comparable ones. Only the United States didn't have that land.

And here's where the whole biofuels calculus breaks down. If the plants used to make those fuels are grown on previously cleared land, then the math behind the carbon lifecycle involved might balance out, with vegetable oil–based biofuels releasing less carbon into the atmosphere than petroleum fuels overall. (The equation incorporates everything from emissions that result from the fertilizers used to grow the crops to those emitted by the trucks used to transport them.) But if in order to grow those crops you must first clear the land of the trees that are now on it, you must also include in your tally whatever carbon was previously sequestered in the biomass and the soils that you've now disturbed. If the land happens to be tropical rainforest, you'll be adding massive amounts of additional carbon. And if it's tropical *peat* forests, your added emissions will be off the charts. Draining a single hectare (2.5 acres) of tropical peat emits an average of 55,000 tons of carbon dioxide every year, roughly the same as burning more than 6,000 gallons of gasoline. The Nairobi-based World Agroforestry Centre has found that peatland-based biodiesels may in fact produce nearly *four times* the emissions of petroleum diesel.

In March of 2019, the European Union, reflecting its more thorough understanding of the carbon lifecycle of palm oil–based biofuels, passed legislation aimed at phasing out their use by 2030. (Despite much back and forth by the Trump administration, the U.S. policy on biofuels remains roughly the same, at least for now.) A few months later, fires again raged across the Indonesian archipelago, torching more than 3,900 square miles of land nationwide—an area the size of Israel—and emitting more than double the amount of carbon dioxide as that year's far-more-publicized Amazon fires. Here in Jambi, they scorched parts of a national park that is home to one of the country's last remaining vast expanses of peat, some of it extending fifty feet into the earth.

The driver of the lime-green truck crouched in the sandy dirt, cigarette dangling from his lips. We'd been scouting out a protected region called Thirty Hills, in southern Riau province, for more than an hour when we'd spotted the guy's vehicle, piled high with illegal fruit, pulled over at the side of the road. Wawan hopped out and began to do his thing. Did the driver know how to get to Tebo? Was this the place to see the tigers? How had the harvest been and where would he be taking his fruit? Wait—his agent was a guy called

Teddy? Wawan had an uncle named Teddy living nearby! Was the guy's Teddy married to so-and-so? Should Wawan have a look at the contract to confirm the odd coincidence? While Dad gabbed, Wari snuck photos of the truck's license plate.

We drove back to the main drag and pulled into a roadside restaurant, eating lunch and then settling in to wait: the driver would have to come this way in order to get to the area's nearest mills. Wari stationed himself on a bench out front and kept his eyes trained on the road, determined not to lose our quarry amid the motorbikes and trucks zooming by. "Hey! Wari!" Wawan shouted at one point, "don't fall asleep!" It was an inside joke; the boss had done just that some years back, causing him to abandon an investigation in which he'd already invested days.

When, four hours later, the green truck finally went rumbling by, the guys grabbed their phones—and I my binoculars and copy of *Birds of Sumatra*—and we scrambled into the Toyota, our driver slamming his foot on the gas. We blew past the truck, Wari aiming his camera from low over the dashboard, and hightailed it toward the mill in question. We were positioned just to the side of the entrance when the driver rolled through with his illicit haul.

Back at the office, Wari's photos would be combined with GPS coordinates, export data, and complicated chain-of-custody charts in a report that would get sent to the companies sourcing from the mill, including, ultimately, such American brands as Kellogg's, Mars, PepsiCo, and Colgate-Palmolive, and a massive Singapore-based refinery that ships biodiesel to countries around the world. Law enforcement and the environmental ministry would also be briefed, in the hope that those masterminding the destruction might eventually be brought to justice. In recent years, EoF had helped to land six officials in jail, including a former *bupati* and several district heads.

(Wawan regretted the smaller players who'd been caught up in the game, including one helpful security guard, who had lost his job thanks to a previous report, and the friendly guys from the platform, who would likely never see those individual plots that they'd been working toward.)

At the height of the 2019 fires, which made headlines around the world, the Indonesian government further stepped up its efforts to contain the damage. President Widodo made permanent a three-year moratorium on licenses for oil-palm plantations, banning the clearing of forests even if they had previously been zoned for "other use," and he announced that the environmental ministry would sue some of the companies responsible for the fires. But he also refused to publish data about, and maps of, oil-palm plantations, citing national security and privacy concerns, in defiance of a 2017 ruling by the country's highest court. In anticipation of the reduced exports to the EU, his government began instituting aggressive biofuels mandates of its own. Having introduced a 20 percent blending rule for domestic fuels in 2016, he now raised it to 40 percent, and added another mandate for power generation from palm oil. Malaysia, too, introduced a mandate to manufacture biofuels with a 20 percent palm oil component, with a plan to raise it to 30 percent soon after. In addition, the two countries moved to expand their biofuels exports, in particular to India and China, and to grow their biofuels processing capacities. And they began eyeing the aviation industry as another potentially significant buyer for palm oil–based fuels.

Despite the now widely acknowledged negative climate implications of such fuels, experts believe that in the coming decade, the majority of growth in global consumption of palm oil will likely be for biofuels. A 2020 report published by Rainforest Foundation

Norway predicted that the result of such expansion could be as many as 13 million acres of additional forest loss—nearly twice the size of Belgium—including some 7 million acres of peatland. In its worst-case scenario, the report predicted that land-use changes related to palm oil–based biofuels could result in some 9 billion metric tons of carbon dioxide emissions. Combined with the smaller but still significant emissions likely to result from new soy oil–based biofuels, annual emissions from such "green fuels" could rival those resulting from the fossil fuels burned by China over the course of an entire year.

And so the industry blazes ahead, expanding to Indonesia's other islands—"The next threat is Papua," Jojo told me. "I went there and—*whoa*. They're cutting down all the trees"—and, as we know, overseas to Latin America and Africa, the latter home to a massive deposit of peat situated at precisely the latitude best suited to growing oil palm.

PART III

◆

FATE OF A FRUIT

9

Nutella and Other Smears

What kills the skunk is the publicity it gives itself.

—Abraham Lincoln

IN LATE November of 2018, a television ad featuring a voiceover by the actor Emma Thompson lit up social media across the United Kingdom. In the animated spot, a baby orangutan swings from the furniture in the sunny bedroom of a young British girl, having fled the destruction of her Bornean-rainforest home. "She throws away my chocolate," we're told of the sad-eyed primate, "and she howls at my shampoo." The ad, produced by a British grocery chain called Iceland with the help of the environmental NGO Greenpeace, never actually aired on TV, having been blocked by Clearcast, the body responsible for approving advertisements in the country.

"You won't see our Christmas advert on TV this year," came a tweet from the Iceland account, "because it was banned." The news of its prohibition—Clearcast had determined that Greenpeace could be considered a political group—sparked widespread debate online, and within a few days the ad had been viewed more than 30 million times. James Corden and other high-profile figures aired their support for its environmental message, and more than 700,000 people signed a petition calling on Clearcast to allow the spot to appear on TV.

Six months earlier, another video, this one featuring a voiceover by Richard Walker, the managing director of Iceland, had also caused a stir. In that piece, the thirty-seven-year-old executive is shown on a research trip in Borneo, descending the steps of a regional airplane, sunglasses perched on his head, and clambering over fallen logs in a dense rainforest. "We went expecting to see some palm oil plantations and some forest clearance," Walker explains, "but nothing prepared me for the industrial scale of what we saw. We witnessed an environmental disaster zone." The video concluded with the announcement that Iceland, whose nine-hundred-plus stores specialize in high-end frozen foods, would eliminate palm oil from its in-house products by the end of 2018.

The backlash was immediate. A snarky video "prepared by Malaysian small farmers" began appearing in the feeds of U.K. Twitter users. Featuring horror-movie music and captions with random capitalization, the ad mocks Walker, dismissing him as "Trust Fund Richard"—his father, Malcolm, who launched Iceland in 1970, is a millionaire—and contrasting him with the hardworking oil-palm farmers of Malaysia. "Multimillionaire Richard took a jet to Borneo," it says, over footage from that original Iceland video, "to lecture Malaysians about the environment."

The Iceland contretemps was the latest battle in an ugly PR war that's now been raging for half a century—ever since palm began seriously threatening other vegetable oils on the global market. In 1986, as Malaysian palm oil started making inroads in the United States—between 1981 and 1985, consumption there increased by

Iceland grocery chain's television ad about palm oil, banned from airing in the United Kingdom.

158 percent—the American Soybean Association, or ASA, responded with a media campaign branding the rival, in the words of one of its full-page newspaper ads, an "unhealthy tropical grease." The group lobbied food manufacturers and the Food and Drug Administration (FDA) to begin labeling "the tropical oils"—coconut as well as palm—as "saturated fats," in hopes of scaring off consumers concerned about possible links with heart disease. "Meet the Man Who's Trying to Put You Out of Business," read one ASA pamphlet published at the time. It featured, in the words of the *Wall Street Journal*, "a surly-looking tropical fat cat," sporting a white suit and wide-brimmed hat and wielding a cigar and coconut drink, sitting beside a barrel stamped with the words "PALM OIL." (The image prompted protests in front of the U.S. embassy in Kuala Lumpur.)

Dan Glickman, then a Democratic representative from Kansas, teamed with Ron Wyden, a Democrat from Oregon, to introduce a bill that would have required food labels to indicate the specific types of oils used in products—at the time they listed simply "vegetable oil"—as well as their quantities of fat and saturated fat. "The unsuspecting consumer," Glickman said, "accustomed to thinking of all vegetable oils as healthful, is easily, and probably intentionally, misled by this kind of labeling." At the time, Glickman was chairman of the Agriculture Subcommittee on Wheat, Soybeans, and Feedgrains.

A year later, the Malaysian industry hit back, launching its own campaign with the help of the American public relations firm Hill and Knowlton. B. Bek-Nielsen, then the managing director of United Plantations (and dad of current CEO Carl), headlined a media tour with stops in cities across the country. Appearing alongside a German-born food chemist named Kurt Berger, a consultant to the Palm Oil Research Institute of Malaysia, or PORIM (precursor to the Malaysian Palm Oil Council), Bek-Nielsen touted the health benefits of the commodity to groups of academics, doctors, manufacturers, and journalists. PORIM also financed research at the University of Wisconsin and elsewhere into ways that palm oil promoted health, publishing reports showing its high levels of vitamin E and beta-carotene, and asserting, in what would become a common refrain, that the whole tropical oils controversy amounted to "a trade issue under the guise of a health issue." Together with the governments of Indonesia and the Philippines (the latter a major producer of coconut oil), PORIM successfully petitioned the Reagan administration to persuade the FDA to reject Glickman's proposed labeling requirements.

But other adversaries emerged. Two years after the launch of the ASA's campaign, an Omaha multi-millionaire named Philip Sokolof kicked off his own vendetta against the tropical oils. Having suffered a heart attack while still in his forties, Sokolof took out a full-page ad in the *New York Times* in which he blamed "the food processors"

Full-page ad that ran in the *New York Times* in 1988.

for endangering the American public by using palm and coconut oils in their packaged foods. "The Poisoning of America!" shouted the headline, printed above a lineup of such familiar pantry staples as

Crisco shortening, Cracklin' Oat Bran, Triscuits, and Pepperidge Farm Goldfish.

By 1989, PORIM was running its own ads in the American press, asserting (correctly) that palm oil did "not require artificial hardening or hydrogenation," which "seems to promote saturation and creates trans-fatty acids," and pointing out that some 70 percent of the soybean oil consumed in the United States was hydrogenated. The Malaysian government and plantation companies commissioned a handful of nutritional scientists who, over the next fifteen years, published numerous studies finding that the kinds of fatty acids in palm oil had a benign effect on blood cholesterol. Prominent among them were Pramod Khosla, a professor of nutrition and food science at Wayne State University, in Detroit, Michigan, and K.C. Hayes, a biology professor at Brandeis University, in Massachusetts. "Financial support for this study was provided by the Malaysian Palm Oil Board," reads the fine print at the bottom of a 2007 study about fats in *Nutrition & Metabolism*. A study on palm oil that appeared in the *Journal of the American College of Nutrition* three years later included the disclaimer, "P.K. is on the speakers bureau for the Malaysian Palm Oil Council and the American Palm Oil Council and has previously received (1996–2005) research support from the Malaysian Palm Oil Board." Nonetheless, the studies circulated widely and were frequently referenced.

But the damage had been done. "American consumers and their health is our concern, and they are telling us they don't want [tropical oils]," a spokesman for the Keebler Company told the *Times* in 1989. "We are getting piles of mail every day, from everywhere." By that time, Keebler, General Mills, Quaker Oats, Pepperidge Farm, Pillsbury, and others had vowed to eliminate tropical oils from their products. The replacement for those 2-billion-some pounds of oil then lurking on U.S. grocery-store shelves? Hydrogenated oils, mostly soy. And though it took a while—thanks largely to the strong-armed tactics of yet another lobbying group, the Institute of Shortening and Edible Oils—the negative health impacts of the trans fats that resulted from hydrogenation eventually became undeniable. (As explained in Chapter 7, the FDA announced labeling requirements for trans fats in 2006 and eventually called for their elimination from the food supply altogether.) Back to the tropical oils!

As the Malaysian and Indonesian economies became increasingly dependent on palm oil, the governments' efforts to defend the commodity intensified. (Malaysia, being far more reliant on exports, tended to lead the initiatives.) Often, the campaigns pushed ethical boundaries. In 2010, twelve of the world's leading scientists, including the former director of the Royal Botanic Gardens at Kew and the biodiversity adviser to the president of the World Bank, became so incensed by the misinformation being promulgated by Southeast Asia's palm oil and logging interests that they wrote an open letter to *The Guardian* and other prominent news outlets. In it, they accused a Melbourne-based consultancy called International Trade Strategies Global, or ITS, of "distortions, misrepresentations, or misinterpretations of fact" in its writings about rainforests, logging, and oil-palm plantations.

While ITS claimed to be independent, the scientists pointed out its close association with a handful of politically conservative think tanks based in Washington, DC, notably the American Enterprise Institute, the Competitive Enterprise Institute, and the Heritage Foundation. ITS's managing director, a former Australian diplomat named Alan Oxley, often lobbied on behalf of Asia Pulp & Paper, a subsidiary of the Jakarta-based conglomerate Sinar Mas and sister company of Golden Agri-Resources, or GAR, one of the largest palm oil businesses in the world (and parent of Golden Veroleum, one of the companies responsible for the land clearing I'd seen in Liberia). Oxley also ran a DC-based nonprofit called World Growth International (WGI), which professed to promote free trade and globalization but tended to focus on opposing environmental regulations, including with attacks on such NGOs as Greenpeace, the Rainforest Action Network, and the WWF for their analyses of deforestation in Southeast Asia.

Both ITS and WGI, the scientists wrote, routinely issued reports that "dismissed or downplayed" important environmental concerns, "including the serious impact of tropical peatland destruction on greenhouse gas emissions and the impact of forest disruption on threatened species such as orangutans and Sumatran tigers." In other cases, they said, the lobbyists' arguments amounted to a "muddying of the waters" designed to defend the credibility of corporations, including Asia Pulp & Paper, that "are playing a major role globally in the rapid destruction of tropical forests."

Malaysia's Sime Darby, the world's largest oil-palm plantation company by planted area, also was applying some interesting tactics at the time. In 2009, a London-based outfit called FBC Media had presented a proposal to Yusof Basiron, then the CEO of the Malaysian Palm Oil Council (MPOC) and a Sime Darby board member, offering to feature him in a documentary that it would then "place" on a reputable news channel "such as BBC World, Bloomberg, CNBC, or Channel News Asia." *The Independent* later reported that FBC had been paid $21 million by the Malaysian government and companies including Sime Darby to develop a "global strategic communications campaign" aimed at convincing "more than 400 million viewers" to support the palm oil industry. FBC promised to include

interviews with other key MPOC figures, "complemented by supporting interviews with . . . industry leaders and Western third-party
champions," and to "put a particular focus on small farm holders"
in order to minimize the idea that the industry was dominated by
"large corporate interests." (Among the FBC's "third-party champions" was the Columbia University economist Jeffrey Sachs. A former
special adviser to U.N. secretary general Ban-Ki Moon, Sachs had
been "cultivated" by the FBC on behalf of Sime Darby, which had
provided a $500,000 grant to Sachs' Earth Institute in 2010.)

An FBC-produced documentary that appeared on the BBC in
early 2009 opened with a cartoon that drew a comparison between
colonialists who asked, "Are the natives friendly?" and modern-day
environmentalists who asked, "Are the natives eco-friendly?" It
was later determined that the company had produced multiple programs for news outfits into which it had inserted propaganda for its
commercial clients, among them Nursultan Nazarbayev, the longtime dictator of Kazakhstan, and former Egyptian president Hosni
Mubarak. When the scandal was exposed in 2011, the BBC and others canceled their contracts with FBC, and the company soon shut
its doors for good.

In the years during which palm oil imports to Europe were steadily
rising, so too were the PR scandals erupting around them. In 2015,
following reports of the commodity's negative impact on the environment and human health, more than 160,000 Italians signed a "Stop
the Invasion of Palm Oil" petition urging the nation's parliament
to introduce a motion banning the substance from public cafeterias.
In response, the newly established Italian Union for Sustainable
Palm Oil (whose members include Unilever, Nestlé, Ferrero, and
trade groups like the Italian Association of Confectionery and Pasta

Cereal box from a grocery store in Barcelona, Spain.

Industries) launched the biggest association-run campaign the country had seen in three decades. For thirty days, Italians could scarcely open a newspaper or turn on the television or radio without hearing about the purported environmental sustainability and health benefits of palm oil.

In France, the country's ecology minister, Ségolene Royal, called for a boycott of Nutella on the grounds that the spread contained unsustainable palm oil. Ferrero responded by pointing out that all of the company's oil was certified by an industry watchdog called the Roundtable on Sustainable Palm Oil (more about that in Chapter 10), prompting an apology from Royal.

At the end of 2015, however, the European Food Safety Authority released a report confirming findings from the Italian health ministry that commercial versions of palm oil likely contained carcinogenic pollutants—a result of processing at high temperatures—prompting companies across the continent to begin adding "palm oil free" labels to their products, including on items that had never contained the substance to begin with.

In June of 2016, France's National Assembly dropped plans for a tax on palm oil—a move it had been considering since 2012—after an outcry from Indonesia and Malaysia, which called the measure a violation of World Trade Organization rules. The Indonesian government had also made it clear that passage of the law might result in the execution of a French citizen then being held in Jakarta on drug-trafficking charges. "We are legislating with a knife at our throats," said one French politician. "The Parliament is being blackmailed."

Later that year, following an international conference on peatlands held in Kuala Lumpur, both the *Borneo Post* and the *Jakarta Post* published stories asserting that oil-palm plantations developed on drained peatland were environmentally sustainable. The articles quoted Malaysian officials who dismissed contrary evidence as propaganda put forth by "militant environmentalists" and "green NGOs." In response, 139 scientists from 115 institutions across the globe submitted a letter in which they condemned the "misleading newspaper headlines and statements" and clarified their position regarding development on tropical peatlands: namely that all current methods of drainage-based agriculture on peat resulted in high rates of carbon loss and were thus environmentally unsustainable.

The 2018 Richard Walker campaign presented as more of the same, featuring the now-familiar talking points about "eco-colonialists" and trade wars masquerading as health and/or environmental concerns that the MPOC had by that point been trotting out for decades. "Richard is Taking from the Poor Farmers in Africa & Asia," concluded the Twitter ad, "to Give to the Rich Western Agri-Corporates that sell Rapeseed & Sunflower."

In fact, in its ham-handedness (as in its random capitalization), the Twitter video bore a striking resemblance to another short spot that had surfaced on YouTube some years earlier. "Al Gore's Penguin Army," which originally aired in 2006, around the time that the film *An Inconvenient Truth* premiered, was a spoof of the former vice president and his environmental concerns. In the two-minute video, Gore is seen boring a group of penguins with a lecture and blaming climate change for every woe confronting humanity. The spot's crude graphics and juvenile sensibility suggested an amateur production, and the person responsible for posting it described himself as a twenty-nine-year-old from Beverly Hills. But a reporter for the *Wall Street Journal* determined that its creator had used a computer registered to a Washington, DC–based lobbying firm called the DCI Group.

DCI, which had close ties to the George W. Bush administration and to various Tea Party figures, often worked with Australia's Oxley. The firm specializes in so-called Astroturf campaigns—corporate efforts disguised to come off as grassroots movements—and among its clients at the time that the fossil-fuel-friendly Gore video circulated were ExxonMobil and General Motors. Previously, DCI had lobbied on behalf of Altria (parent company of Philip Morris USA) and the former Burmese military junta. From 2000 to 2006, it had published an online magazine, *Tech Central Station*, that described itself as a website "where free markets meet technology" and featured articles questioning the reality of climate change. Oxley served as a writer and editor for its Asia-Pacific pages.

Soon after the Richard Walker Twitter spot began appearing, news outlets reported that it, too, had been the work of DCI Group. The video had been promoted by an organization called the Human Faces of Palm Oil, which, according to its website, "seeks to advocate on behalf of Malaysian small farmers" and is a joint project of MPOC, FELDA, and something called NASH, the National Association of Small Holders, among others. In an August 2018 proposal to MPOC and other industry groups, DCI had recommended that independent farmers be the "primary messengers" globally of the campaign against the industry's critics, with Human Faces of Palm Oil leading the campaign. The proposal made clear that DCI had been coordinating other grassroots campaigns through industry front groups, including something called Palm Oil Farmers Unite. The Farmers Unite Facebook page features a matrix of *National Geographic*–esque headshots of farmers from across the globe and says that the organization "speaks on behalf of 7 million oil palm smallholders" fighting "dangerous campaigns and policies that threaten our livelihoods."

In the 2018 proposal, DCI also pitched the creation of an African platform with the help of a Nigerian think tank, the Initiative for Public Policy Analysis. "Recruiting support from African allies is necessary to put maximum pressure on your opponents," came the pitch from the Beltway strategists. European NGOs and politicians, it added, "fear accusations of neo-colonialism and discrimination." That year, Farmers Unite and NASH also took out several ads in *Politico Europe*, an outlet read widely by legislators in Brussels (where DCI also has an office), denouncing the EU's palm oil policy as "crop apartheid."

At a meeting in Kuala Lumpur with MPOC and other industry leaders in early 2019, DCI presented an additional proposal in which it doubled down on its Astroturf strategy. "Attempting to reason with these opponents, through dialogue or scientific research, will not stop their attacks and will not advance Malaysia's position," it said. "Small farmers are Malaysia's most powerful weapon against Europe and the NGOs."

In contrast to the media campaigns, smallholder organizations in Indonesia and Malaysia have long complained that their governments support large palm oil corporations and leave their members struggling at the poverty line. The average palm oil worker in Indonesia is likely to make about $6 a day. Malaysia's daily wage for palm oil laborers is closer to $9. The men who run the industry, on the other hand, rank among the wealthiest in all of Southeast Asia. Major Wilmar investor Robert Kuok, whose net worth is $11 billion, is the richest man in Malaysia. The company's CEO, Kuok Khoon Hong (a nephew of Robert), is worth $3.6 billion. The Salim Group's Anthoni Salim has a fortune estimated at $5.5 billion, and when Sinar Mas founder Eka Tjipta Widjaja died in 2019, he was worth $9.1 billion.

The resumés of the industry's overseas ambassadors also raise some questions. In his Amazon biography, "Dr. Bruce Fife, C.N., N.D.," who wrote the 2007 book *The Palm Oil Miracle*, describes himself as the author of more than twenty books and a "speaker, certified nutritionist, and naturopathic physician." Reached at the Colorado Springs offices of Piccadilly Books, which published *The Palm Oil Miracle*, Fife told me that the initials after his name refer to "certified nutritionist" and "doctor of naturopathy," adding that he'd earned his degrees from the Clayton School of Natural Healing. Having never heard of Clayton, I inquired as to its location, a question that was met by a long pause. "In Georgia," Fife finally said. In fact, Clayton College, an unaccredited correspondence school, was located in Birmingham, Alabama. It closed abruptly in 2010, just as a new law went into effect requiring that degree-granting institutions in the state be approved by an agency recognized by the Department of Education. In the past, Clayton's course offerings had included instruction in aromatherapy, Bach Flower remedies, therapeutic touch, and psychodietetics. Among its graduates were Hulda Clark, who has said that all cancers, as well as AIDS, are caused by "pollutants and/or parasites," and that she could cure them with herbs and a low-voltage electrical device; and Robert O. Young, who in 2014 pleaded guilty to multiple counts of grand theft and conspiring to practice medicine without a license. In 2011, Clayton settled a class-action lawsuit with some fourteen thousand former students, agreeing to pay $2.31 million in reimbursement for tuition that it had failed to refund after shutting down.

In January of 2019, the World Health Organization published an article in which it called for greater scrutiny of the health and planetary impacts of palm oil and likened the industry's tactics—"establishing lobbying structures in political and economic hubs, fighting regulations, attempting to undermine reliable sources of information, and using poverty alleviation arguments"—to those previously deployed by the tobacco and alcohol lobbies. Writing in the *Bulletin of the World Health Organization*, Sowmya Kadandale, of the United Nations Children's Fund; Robert Marten, of London's School of Hygiene & Tropical Medicine; and Richard Smith, of the University of Exeter's College of Medicine and Health, warned of a "cocktail effect," in which palm oil could be damaging to health when combined with other ingredients used in highly processed foods. The authors added that the food industry's marketing of "ultra-processed" products to children also was reminiscent of the way the tobacco and alcohol industries had targeted young people.

The report called the palm oil industry "an overlooked actor in discussions on non-communicable diseases" and suggested that policy-makers consider ways to reduce the demand for oils and for unhealthy ultra-processed foods. It also advised avoiding "the influence of lobbying by food industries whose practices adversely impact human and planetary health." The authors called on academics to "exercise caution when engaging in research activities using funding from the palm oil and related industries."

In response to the report, a spokesman for the Malaysian Palm

Oil Council released a statement objecting to the fact that none of its authors were "palm oil scientists" and adding that it was unclear whether any of the report's reviewers had been "palm oil scientists" either, thus suggesting a potential bias. "The authors conveniently ignored key palm oil publications in respected journals," wrote Kalyana Sundram, current CEO of MPOC, "and cherry-picked a handful that fitted their hypothesis."

Having been issued just six weeks before the DCI Group presented its follow-up proposal to MPOC in Kuala Lumpur, the WHO report may have provided the lobbying group with some of its talking points. "The eco-colonialists have turned to a scorched earth approach of junk science and faulty logic," DCI wrote in its pitch. "They label palm oil as the new tobacco." The group's proposal involved a budget of more than $1 million.

It wasn't the first time that the industry had clashed with the WHO. In 2003, Derek Yach, a South African doctor who had formerly served as executive director of non-communicable diseases at the agency, co-authored a report on diet, nutrition, and prevention of chronic diseases co-sponsored by the UN's Food and Agriculture Organization. It included, Yach said, "a few lines" about how the saturated fat from palm oil threatened cardiovascular health, and recommended a reduction in global consumption of the oil. After the committee put the report out for public review, Yach said, the Malaysian ambassador to the U.N. showed up at his Geneva office with a delegation and argued that any effort to curb consumption of the commodity would threaten the livelihoods of several million people. She demanded that Yach and his colleagues change the text. His experience with the palm oil lobby, Yach added, was "substantially worse than anything I ever faced from the tobacco sector."

· · ·

More recently, as the European Union has deliberated over the final shape of its biofuels legislation, the lobbying battle has reached a fever pitch. Indonesia and Malaysia have threatened to restrict European imports and to undertake other trade reprisals. The new Malaysian prime minister, Mahathir Mohamad, sent a letter to French president Emmanuel Macron suggesting that he would suspend trade talks and be forced to inflict "regrettable economic and trade consequences" for £6 billion ($6.5 billion) worth of French exports as a result of the "de facto ban" on palm oil. (The legislation actually allows for the import of palm oil produced by smallholder farmers, cultivated on unused or severely degraded land, and derived from improved yields to be counted as renewable energy.) The country's primary industries minister, Teresa Kok, decreed the biofuels decision "discriminatory against the economies of developing nations in Southeast Asia, Africa and Latin America" and said it was "designed to hurt the livelihoods of millions of small farmers."

In May of 2019, the British department store Selfridges followed Iceland's lead, announcing that it had eliminated all palm oil from its high-end Selfridges Selection range, a total of 280 products. Four months later, Malaysia's biggest supermarket chain announced that it would no longer sell products marked "palm oil free," and the government said it was considering whether to ban such products altogether. Indonesia's food regulatory agency banned food labeled "palm-oil free" in August, and by early 2020, Yusof Basiron, who is now executive director of the Council of Palm Oil Producing Countries, was calling NGOs critical of the commodity "toxic entities." Addressing the audience at an industry forum in Kuala Lumpur, Franki Anthony Dass, the chief adviser and value officer for Sime Darby, referred to the groups orchestrating attacks on palm oil and asked, "If they are so unfriendly, why allow them to be in our countries Malaysia and

Indonesia? We have the right to control this and do something drastic for once."

Just a few weeks earlier, Philip Jacobson, an American journalist
who has written numerous stories about environmental destruction
linked to the palm oil industry, had been deported from Indonesia
after spending several days in a Kalimantan prison for a purported
violation of his visa. Soon after, an Indonesian journalist who'd published an article about indigenous Dayak communities protesting a
palm oil company also was arrested. The operation in question was
a sister company of the one that Muhammad Yusuf, the now-dead
journalist mentioned in Chapter 8, had covered a year earlier. If anything, the pressure was only intensifying.

10

Fight the Power

We can't save the world by playing by the rules.
Because the rules have to be changed.

—climate activist Greta Thunberg

IN THE minutes before dawn on November 17, 2018, the international waters off of Spain's Gulf of Cadiz were relatively calm. Not so the deck of the 237-foot *Esperanza*, whose green-painted hull and giant dove-and-rainbow logo would have been just discernable in the near dark. There, a group of Greenpeace activists from a dozen countries were shouting over the roar of the engine as they scrambled to pull on splash-proof overalls and helmets and gather backpacks and climbing gear. Six of them jumped from the ship into three motorized RIBs, or rigid inflatable boats, whose drivers then sped off in the direction of a tanker looming on the horizon.

Laden with palm oil from a Sumatran refinery owned by Wilmar International, the 607-foot *Stolt Tenacity* was making its way toward the Dutch port of Rotterdam when the first Greenpeace RIB pulled stealthily up to its side. The team's leader, a British Columbia native named Victoria Henry, spooled out a steel caving ladder as a deck-hand heaved a long pole in the direction of the *Stolt*'s deck, catching its hooked end on the rail. He pulled down hard, disengaging the pole and leaving the ladder dangling down the side of the tanker. Handing the pole off to a colleague, Henry jumped to the ladder and hoisted herself up the side of the fast-moving ship. The second RIB pulled in just close enough for the next climber to make the leap from wobbly boat to dripping ladder, then sped away to clear an ocean landing should he lose his footing.

Just as the sixth activist was pulling her gear over the rail of the *Stolt*, a light flicked on up on the bridge. Minutes later, the ship's captain came stomping toward the group, barking at his crewmen to confiscate their belongings and waving his arms wildly. "You're pirates!" he raged, whacking the Indonesian activist on her helmet. "Get the hell off my ship! I'll throw you overboard!"

The plan had been for the climbers to quickly make their way toward the front of the tanker and there set up a makeshift camp. They'd come equipped with sleeping gear and enough food and water to last the duration to Rotterdam, which they'd assumed would be three or four days. A Greenpeace photographer and videographer had documented the early-morning boarding from their vantage points on the RIBs and were now headed back to the *Esperanza*, where the captain had radioed the *Stolt* to inform its captain that his tanker was being boarded by peaceful, unarmed protesters. "We weren't against his ship," Henry told me by phone from her home in London, "but against his cargo."

Two months earlier, Greenpeace International had published a report finding that Wilmar, which trades some 40 percent of the world's palm oil, was getting some of its supply from eighteen mills linked to deforestation, in obvious violation of its 2013 NDPE policy. The Gulf of Cadiz boarding had come on the heels of another

Boarding the *Stolt Tenacity* in the waters off of Spain.

"action," carried out at a Wilmar refinery on the Indonesian island of Sulawesi, in which a group of Greenpeace volunteers had scaled a storage tank and painted "DIRTY" in fifteen-foot-tall letters. A second group had climbed the anchor chain of the tanker set to transport the oil, thus impeding its departure.

The *Stolt* activists were herded into the six-bed "Suez cabin." (Such rooms were designed to accommodate the handful of armed guards that will often board a ship for a dangerous leg like the one through the Middle Eastern canal, the narrowness of which enables pirates to strike without the aid of an attention-getting mother ship.) They had assumed that the captain would simply continue on his way to the Netherlands—the daily operating costs of a cargo ship can run to the tens of thousands of dollars—but now it appeared that he'd turned around and was steering the ship in circles. As a crewmember stood watch by their door, the activists did their best to determine their location by the angle of the sun, and to avoid imagining the worst. ("What if we end up in a Moroccan prison?") It didn't help that the hotheaded captain kept marching back in to harass and threaten them. Finally, late on the afternoon of November 18, one of the group looked toward the cabin's window and let out a gasp. The others turned to see just as the Rock of Gibraltar slid into view. "'Oh my fucking god,'" Henry remembers thinking, "'we're in Spain.' I will never forget seeing that rock."

Thirty-three hours after they'd originally scaled her side, the *Stolt Tenacity* eased her way into the Spanish port of Algeciras. ("As awful a place as you could muster," Kevin Barry calls the town in his 2019 novel *Night Boat to Tangier.*) Following a meeting with the captain, a lawyer for the shipping company, and local law enforcement officials, Henry led her five colleagues through immigration and out into the Algeciras night. The next morning they drove to Lisbon, where the *Esperanza* was waiting to welcome them back onboard before continuing on to Rotterdam.

My introduction to the world of palm oil activists had come a few years earlier, when I'd flown out to San Francisco to spend some time with folks from the Rainforest Action Network. Founded in 1985 and with a staff of some fifty, RAN often collaborates with Greenpeace and other NGOs in its efforts to push rainforest-threatening industries to clean up their acts. On a bright morning in March, I'd driven across the Bay Bridge and pulled up in front of a low-slung warehouse on an industrial block of West Oakland. Gathered inside the vast, graffiti-splashed bunker, which is owned by Greenpeace, were a handful of RAN staffers and twenty or so volunteers, among them a home-schooled twelve-year-old who'd flown in from Arizona and a soon-to-be-seventy "orangutan fanatic" from Seattle.

The group had been recruited to participate in a four-day kick-off campaign targeting Quaker Oats and its parent company, Pepsi-Co, for using "conflict palm oil" in their popular snacks. RAN had recently launched an initiative, the Snack Food 20, calling on companies including PepsiCo (plus Nestlé, Unilever, Mars, Hershey's, Kellogg's, and others) to wield their purchasing power to demand substantive change from the industry. Some weeks earlier, staffers from the NGO had delivered a report documenting rainforest destruction, land grabbing, and forced labor on PepsiCo-linked plantations to the company's headquarters, in Purchase, New York. The Oakland event was meant as a sort of training operation for a follow-up "week of action" slated for a few months later. The plan was for hundreds of activists to descend on grocery stores and public spaces across the country and inform consumers about Quaker's possible links to tainted oil.

By the time I arrived, the volunteers were bleary-eyed, having stayed up most of the night to rewrite song lyrics, choreograph dance moves, and hand-letter banners in preparation for an action they'd planned for that afternoon. Over sprouted whole-grain English muffins and vegan cream cheese, the variously pierced and tattooed participants plotted the details of the protest, debating who would play the parts of Larry, the white-haired Puritan from the Quaker label, and of Strawberry, the orangutan whose rainforest habitat was under siege. A freelance producer from Minneapolis took to a whiteboard to sketch a map of the interior of a Target store at the corner of Mission and Seventh streets, the venue having been selected based on the previous day's citywide reconnaissance effort. "There is a security camera here," she said, illustrating her point with an X. We moved to an adjacent room, a cavernous space lined by industrial shelving crammed with life jackets, giant orange buoys, and rolled-up sleeping bags, and the group split into two to rehearse its improvised call-and-response.

The discussion turned to the risks inherent in the sort of nonviolent action the group was about to undertake. "So what do you do when a cop comes up to you?" asked the consultant who'd been brought in to lead training for the event. She ran through scenarios involving police officers, security guards, and escalating emotions. "It's normal to get nervous," RAN's thirty-five-year-old executive director, Lindsey Allen, had told the gang the night before. "But what you believe really does matter going into an event like this."

At the time, few American voices had made themselves heard on the subject of palm oil. Among the exceptions had been a pair of determined eleven-year-old Girl Scouts. In 2007, while preparing for a science fair presentation about rainforests in conjunction with earning their Bronze Awards, Ann Arbor natives Rhiannon Tomtishen and Madison Vorva had discovered that the cookies they'd been selling, household names like Thin Mints and Tagalongs, contained palm oil sourced from what had once been orangutan habitat. The two boycotted their annual sale and began to petition Girl Scouts of the USA to stop using the oil in their cookies.

They had achieved only modest success by 2010, when the girls reached out to Glenn Hurowitz, then the managing director of a Washington, DC–based organization called Climate Advisers. A few years earlier, Hurowitz had published an op-ed in the *Los Angeles Times* in which he'd made the connection between palm oil and tropical deforestation. Outspoken on the latter since his days as a member of the Yale Student Environmental Coalition, the New York native had served briefly as a media director for Greenpeace, and he well understood the publicity gold that was two adorable teenagers distraught over the fate of endangered primates. Hurowitz and his colleagues arranged media training for the girls and teamed with RAN and other NGOs to coordinate protests and online petitions highlighting their cause. They focused their efforts on Kellogg's, which at the time was producing Girl Scout cookies (they are now predominantly made by Nutella manufacturer Ferrero) and which has its headquarters in the girls' home state of Michigan. Before long, articles about Tomtishen and Vorva had appeared in the pages of the *New York Times*, the *Wall Street Journal*, and *Time* magazine, and the girls had taken their case directly to the American consumer via the airwaves of NPR, Fox News, and ABC's *The Early Show*. The resulting pressure eventually led Kellogg's to commit to reducing its use of palm oil, and to sourcing whatever it did buy from companies with NDPE policies. ("I think it's time," came the response from Malaysia's plantation minister, Bernard Dompok, "that some of these little girls get better informed on these things.")

By then, the industry had come together to establish its own mechanism for dealing with the mounting criticism. In 2004, pressure from RAN, Greenpeace, and other organizations had led a group of prominent growers, manufacturers, and retailers to establish something called the Roundtable on Sustainable Palm Oil, or RSPO. The organization developed a set of environmental and social criteria that companies could comply with in order to be deemed "sustainable." In 2008, Malaysia's United Plantations became the first company to be certified by the RSPO. UP's CEO, Carl Bek-Nielsen, has served as co-chair of the organization since 2014. Representatives from NGOs such as WWF, the World Resources Institute, and the Forest Peoples Programme sit on the RSPO's board of governors, but twelve of its sixteen members represent palm oil processors, manufacturers, retailers, banks, investors, and food-processing companies—which may explain why the organization's progress has been so limited.

It isn't for a lack of cash. The group's annual meetings, invariably sponsored by industry giants Wilmar, Sime Darby, Musim Mas, PepsiCo, and Cargill, are reliably splashy affairs, with hundreds of members flying in from around the world to enjoy three days of keynote speeches, breakout sessions, buffet lunches, cocktail hours, and long, boozy dinners. Rotating among such Asian metropolises as Jakarta, Kuala Lumpur, and Bangkok, the conferences often are hosted by Shangri-La Hotels, the high-end chain owned by Wilmar investor Robert Kuok.

But many of the people I spoke with for this book told me that RSPO certification too often serves as a fig leaf for companies looking to obscure their unsustainable (and, in some cases, illegal) practices while continuing to secure investment and market access. In 2015, the London-based Environmental Investigation Agency (EIA) released a report criticizing the RSPO's certification process, which involves third-party auditors that are contracted and paid for by the same companies they then evaluate. The report provided evidence that auditors had colluded with plantation companies to disguise environmental and social violations, including instances where management had intimidated workers before auditor visits. "Casual workers" on a plantation in North Sumatra had been told to retreat to remote areas, away from the roads on which the auditors might be driving, and others were instructed to lie about the very existence of such workers. In Honduras, where I visited the plantations of Grupo Jaremar a week after the RSPO auditors had been there, I heard similar stories. "I've been in four audits," one worker told me. "The engineers intimidate us while we're talking to the RSPO, and the next day there are reprisals." Others said that they'd been carefully coached.

"The immediate supervisors have trainings with the workers and say, 'This is what you have to say.' They take the international auditor to the ones who know the scripts." Still others spoke of the free soda and "banquets" enjoyed afterward by those who had performed well.

In 2014, Greenpeace, RAN, and other NGOs got together with some of the RSPO's more progressive industry members to establish something called the High Carbon Stock Approach, a mechanism intended to help plantation companies determine what land is suitable for development and what should be protected, based on its carbon content and biodiversity value. The approach has since proven effective, but its existence hasn't solved the organization's credibility issues. In 2016, the Australian watchdog group Palm Oil Investigations, which had previously worked to persuade companies to join the RSPO, withdrew its support for the organization, declaring that it had lost faith in its ability to rein in the industry. The announcement followed the resignation of a Switzerland-based foundation called PanEco, which works to protect orangutans (including by supporting Ian Singleton's center), over the RSPO's "sheer level of inaction."

In 2018, the RSPO responded to its critics by adopting a new set of "principles and criteria" featuring more stringent prohibitions against deforestation, planting on peatland, and labor and human rights abuses. It also upgraded its much-criticized process for dealing with industry complaints. Less than a year later, however, after sanctioning FELDA asset FGV Holdings Berhad for more than twenty-five breaches of its certification criteria—including the forced labor documented in Chapter 6—the RSPO reinstated the Malaysian company's certification, despite a lack of evidence that it had done anything to address the issues in question. Other companies found in violation of RSPO principles and criteria have taken to quitting the group rather than institute the reforms that it recommends. In 2018,

Liberia's Golden Veroleum (GVL) withdrew its membership from the RSPO after being found by the group to have appropriated community land, leveled forests, and despoiled gravesites. The following year, the EIA issued a second report in which it documented thirty-eight cases that remained open in the RSPO complaints system, one of them for nine and a half years. Thirty percent of the other complaints had been pending for more than three years.

Most telling of all may be the fact that, seventeen years into its existence, the RSPO has currently certified just 19 percent of the global palm oil supply. Only a tiny percentage of consumer goods bear the organization's logo, and most Americans wouldn't recognize it anyway. "I don't think the RSPO has played much or any role in reducing deforestation in the palm oil industry," Glenn Hurowitz told me. "It's provided a green-washing tool for major consumer companies."

The scrappy Girl Scouts campaign, on the other hand, would go on to yield results even years after the fact. In the summer of 2013, by which time Vorva and Tomtishen had graduated from high school, fires were raging across Indonesia, as they would be again in 2015 and 2019. That fall, Greenpeace issued a report in which it traced many of the conflagrations to plantations linked to Wilmar. In an interview that appeared on Bloomberg TV in Asia, Hurowitz called out the Singapore-based trader for its role in the fires and the dangerous haze that they were engendering.

Wilmar's notoriously private CEO, Kuok Khoon Hong, responded by writing Hurowitz a letter. Having recently worked alongside Rainforest Foundation Norway to convince that country's government to drop twenty-three "dirty" palm oil companies, Wilmar among them, from its then–$700 billion pension fund, the American sensed that the timing might be right. He figured that if he could leverage Wilmar's enormous reach, there might be a chance to push the whole industry dramatically forward. Kuok had made it clear that he was interested in improving Wilmar's environmental reputation, and Hurowitz had enlisted the Switzerland-based Forest Trust to speak with the company about how to map out a path toward doing so. But now the executive was getting cold feet. He was worried about taking such a major step without the industry's other big players—namely GAR, Cargill, and Musim Mas—doing so along with him.

In late November, after numerous conversations between the men, and under continued pressure from Greenpeace and others, Kuok invited Hurowitz to join him for a face-to-face meeting. Upon touching down in Singapore, the campaigner sent the CEO a carefully calculated text. Attached to it was a photo of protesters stationed in front of the Kellogg's building in Michigan. (Wilmar had recently signed a joint-venture with the American company.) "Every one of your customers' headquarters is going to look like this," Hurowitz wrote. "This is an opportunity to distinguish yourself."

Over the next forty-eight hours, Hurowitz and his Forest Trust counterpart worked around the clock to hash out a deal with Kuok, who agreed to make a comprehensive no-deforestation pledge under the condition that it lock in a major contract that he'd been hoping to win from Unilever. (The food giant founded by William Lever did its part for the environment by agreeing to demand the provision.) Within a year, most of the other leading palm oil players had made similar commitments to ending deforestation and planting on peatlands, and to avoiding human-rights and labor abuses.

Hurowitz saw the victory as a lesson in the power of consumer voices strategically harnessed. Over many years, he told me, "European pension funds would send their sustainability officers to Southeast Asia to pull politely on the sleeves of palm oil executives and ask them to stop deforestation. The palm oil executives would say, 'Oh yes, this is a terrible problem, and we will work hard to do something about it.' Or, 'We'll sign up to RSPO.' And the pension funds—and it was basically true of the consumer companies as well—that was their pledge: 'Please do something better, we're engaging, we're talking to you.' It didn't produce much action. And because the consumer companies continued buying palm oil, and the institutional investors continued investing even though their exhortations to improve were being ignored, the palm oil industry didn't have to take what these foreigners said seriously."

The photos taken to commemorate the Wilmar deal show three smiling men—Kuok, Hurowitz, and the Forest Trust's Scott Poynton—but the victory had its roots in the efforts of many, many more. Among them are the U.S.- and Europe-based staffs of such NGOs as RAN, Greenpeace, and Friends of the Earth. They may be better known for scaling skyscrapers and chanting in the streets, but the campaigners also draft meticulously sourced reports and drag out their business attire in order to pull up chairs at conference tables and hammer out deforestation plans. And they will invariably tell you that their success depends almost entirely on the work of their counterparts on the ground. All of the NGOs collaborate with grass-roots outfits based in places like Medan and Sandakan, Monrovia and Guatemala City, pooling energy and resources to coordinate fact-finding trips to rural areas and arranging for country nationals to visit places like Washington, DC, and Brussels, where their accounts of the realities they face at home often are at odds with the pretty words offered by executives and trade ministers.

A few years back, one such all-hands-on-deck initiative helped to avert a disaster in the making in Central Africa. In 2009, the government of Cameroon had approved a plan for a New York–based agribusiness called Herakles Farms to develop a 180,000-acre oil-palm plantation in the country's southwest. Herakles had begun razing forest on the periphery of a national park that was home to chimpanzees and endangered drill monkeys, and to leopards, buffalo, and elephants. Cameroonian villagers staged ongoing protests at the site of the project, and, in 2012, scientists from around the world collaborated on an open letter expressing their deep concern for its impact. The large-scale clearing of forest for it and similar plantations, they wrote, would likely lead to the extinctions of already endangered primates. A year later, Greenpeace International and the U.S. based Oakland Institute released an exposé highlighting the many untruths that the company had put forth to investors and local communities in regard to the project's environmental and social prospects. The groups demanded a thorough evaluation of the proposal, which revealed that, among other concerns, parts of the Herakles concession qualified as areas of high conservation value. In 2013, after a rash of bad publicity in the Cameroonian and international press, Herakles finally suspended the project.

Local groups in Cameroon and Sierra Leone have likewise been leading the charge in a years-long battle against Socfin, the Luxembourg-based company founded by Adrien Hallet. Since 2011, the communities have sought reparations for a litany of complaints in regard to its land-acquisition policies and the labor conditions on its plantations. In 2013, a delegation of laborers from Socapalm, the company's Cameroonian subsidiary, traveled to Paris to air their grievances, and a few years later a French television network released a documentary about working conditions on Socfin plantations. In the film, one protester carries a sign reading "We are slaves of Vincent Bolloré," a reference to the billionaire Frenchman who today

Liberian lawyer Alfred Brownell accepting the 2019 Goldman Environmental Prize.

owns a 39-percent share of the company. Bolloré filed a defamation suit against the network, one of some twenty claims that he'd made against journalists, lawyers, and NGOs since 2009. (The network won the case.) Eventually, a collective formed in response to the mogul's continued legal threats: On Ne Se Taira Pas—We Will Not Shut Up.

In the early 2000s, Alfred Brownell, the Liberian lawyer I mentioned in the prologue, collaborated with international NGOs including Greenpeace and the London-based corruption watchdog Global Witness to track the timber that was then fueling Charles Taylor's civil war in the country. By lobbying for U.N. sanctions, the coalition ultimately helped bring the bloody conflict to an end. More recently, Brownell and village leaders had teamed with Friends of the Earth to confront Sime Darby and GVL about their operations in Liberia. Brownell, who ran his law firm, Green Advocates, out of a third-floor walk-up in downtown Monrovia, eventually determined that Sime Darby had failed to abide by various RSPO principles, including in regard to securing community buy-in and avoiding areas of high-carbon value, and made the objections public. President Ellen Johnson Sirleaf became so angered by his interference in her efforts to secure foreign investment that she intervened with him personally. The ongoing harassment he received, which included at least one attempt on his life, eventually led Brownell to flee the country. Today, he serves as a human rights fellow at Yale University's law school, and in 2019 he received the Goldman Environmental Prize, the same honor that Sumatra's Rudi Putra had been awarded a few years earlier. By January of 2020, Brownell and his collaborators had put so much pressure on Sime Darby that the company had pulled out of Liberia altogether. "They tried to paste the Southeast Asian model [of oil-palm plantations] into Liberia," he told me, "and it didn't work."

Though Brownell managed to escape with his life, many who have stood up to the global palm oil industry haven't been as lucky. Over the years, as plantation companies have swallowed up ever greater swathes of the planet's fertile land, incidences of violence against those who would stand in their path have continued to mount. In 2020, Global Witness found that 212 land and environmental defenders had been reported killed in 2019—an average of more than four a week. The agribusiness sector, palm oil interests prominent within it, ranked second only to mining as the deadliest focus for activists.

In 2015, a Thai campaigner named Chai Bunthonglek was shot dead after having worked on behalf of an indigenous community trying to win a legal land battle with a local palm oil company. He was the fourth activist campaigning against the development to be killed in five years. A year later, a twenty-eight-year-old Guatemalan schoolteacher named Rigoberto Lima Choc was killed on the steps of a courthouse in the northern city of Sayaxché. Choc had led a group of activists that had filed a criminal complaint against a palm oil company based on evidence that its overflowing effluent ponds had triggered a large fish kill along a sixty-five-mile stretch of a local river. Less than a year after that, an activist named Bill Kayong was shot and killed while sitting in his pickup truck in the Malaysian city of Miri. Kayong had been working with a group of villagers who were trying to reclaim land that the local government had transferred to a Malaysian palm oil company.

In 2012, a Honduran human-rights lawyer named Antonio Trejo Cabrera was ambushed and killed by gunmen while walking out of a church in the capital, Tegucigalpa. Trejo had been representing local peasant organizations in a fight against the palm oil giant Grupo Dinant, and had recently won a handful of cases, forcing the company's plantations to be turned over to local residents. In 2013 the Office of the Compliance Adviser/Ombudsman of the World Bank's International Finance Corporation investigated Grupo Dinant and, citing allegations that forty different murders could be linked to its plantations, security guards, and third-party security contractors, declined to provide the company with a multimillion-dollar loan installment. Most of the victims in the cases cited by the IFC were local activists and farmers.

The situation continues to deteriorate in Indonesia. In November of 2019, two Medan-based journalists who had written pieces critical of the industry were found dead with stab wounds on an illegal oil-palm plantation in Sumatra. Maraden Sianipar and Martua Siregar had been reporting on the company behind the plantation and on a community group trying to gain control of the land after authorities ruled that it had been cleared illegally. The deaths occurred just a month after an activist named Golfrid Siregar was found unconscious with severe head injuries on a traffic overpass in Medan. Siregar, who died three days later, was known for his advocacy work with communities ensnared in land conflicts with palm oil companies. Police ruled his death the result of a drunken-driving accident, but his former colleagues disputed the claim, pointing to several holes in the evidence, including the fact that Siregar wasn't a drinker.

If nothing else, the running tally of arrests and murders tied to the industry, particularly in Southeast Asia, offers a window into the difficulty involved in attempting to rein in a sector so critical to the regional economies and intimately linked to the halls of power. Among the prominent actors still facing intense scrutiny is Sinar Mas, the company long connected to climate-change denier Alan Oxley. In 2019, Greenpeace issued a report finding that between 2015 and 2018, an area larger than Singapore had burned in a concession linked to Sinar Mas and its subsidiary APP. As recently as March of 2020, its sister company, Golden Agri-Resources, whose buyers include Nestlé, Procter & Gamble, and Unilever, was credibly accused of illegally operating oil-palm plantations inside forest zones protected under Indonesian law.

Indofood, too—founded by a Suharto crony named Anthoni Salim—continues to be called out for a range of offenses. For many years, the Jakarta-based food processor has enjoyed a joint venture with PepsiCo, making the American giant's branded products in Indonesia. Indofood sources much of its palm oil from Indofood Agri Resources, or IndoAgri, the plantation arm of the Salim Group. Though PepsiCo is among those to have signed an NDPE, its agreement featured a loophole under which third-party suppliers such as Indofood were exempt. Over the years, RAN and others have documented a series of RSPO violations, including labor abuses and peatland clearance, on IndoAgri's plantations. In 2016, activists scaled the six-story Pepsi landmark in Long Island City, Queens, and dropped a one-hundred-foot "Cut Conflict Palm Oil" banner to call attention to the company's compromised deal with Indofood. Two years later, the RSPO found IndoAgri in breach of its standards and set a deadline for the company to finally address the violations. Instead, in February of 2019, IndoAgri withdrew its membership from the organization.

Having worked with traders and plantation companies over many years, Glenn Hurowitz understands the political landscape in Southeast Asia, and he suggests that executives often may want to do the right thing but feel they lack the political cover to make it happen. In order to negotiate such realities, and to hold the industry to account

An April 2016 protest in Long Island City, Queens, organized by the Rainforest Action Network.

more generally, Mighty Earth launched something called Rapid Response, a satellite-based monitoring system designed to quickly identify deforestation and new peatland development across Indonesia and Malaysia. These days, the program keeps tabs on some 51 million acres, more than twice the size of Portugal, and publishes a monthly report in which it identifies the deforestation taking place (related to soy and cattle production in addition to oil palm) and the plantations and buyers linked to it. The organization then files grievances with the major traders, who are obligated to contact offending companies in their supply chains within forty-eight hours. If, after two weeks, the non-compliant actors have failed to stop chopping down forests, they are cut off by the traders, thereby losing access to the global market.

Increasingly, opponents of dirty palm oil, like campaigners against fossil fuels, are focusing their efforts on the industry's financial backers. In April of 2019, before changing into his traditional African tunic for the Goldman Awards ceremony, Alfred Brownell pulled on a bright-red T-shirt with the words "BlackRock: Stop Making the Climate Crisis Worse" and parked himself in front of the global asset manager's San Francisco headquarters. The company is the largest American investor in palm oil, managing nearly $600 million in the commodity, including through investments in Golden Agri-Resources and Sime Darby. At the time of Brownell's protest, the two controlled a combined 1.5 million acres in Liberia. (BlackRock also funds Indofood parent the Salim Group.) The fund's CEO, Larry Fink, had recently made a public statement about how important it was for businesses to have a "social purpose," and Brownell intended to call him out on it. "If this is what you believe in," he said of the billionaire, "then why are you investing in Golden Agri-Resources, which is carrying on this annihilation in my country?"

· · ·

In the aftermath of the Greenpeace action in the Gulf of Cadiz, the trade press covering the international shipping industry ran stories about the incident, warning that tankers transporting palm oil would need to pay closer attention to security. Wilmar had been the subject of negative publicity elsewhere, including during protests directed toward such clients as Nestlé, Unilever, and Mondelez for their purchases of the trader's dirty palm oil. In December of 2018, less than a month after the ocean boarding, the company announced a new commitment to better enforce its policies, including by mapping and monitoring its suppliers, and by cutting off those who continued to destroy forests.

"It's people around the world who have made a difference," Ian Singleton told me that day in his office. He recounted how, a few years earlier, young activists in Indonesia had pooled their talents and resources to push the government to finally clamp down on the powers behind the recurrent forest fires. They had put together online petitions and arranged it so that each time somebody signed on to them, emails got sent out to various politicians, including those linked to Norway's $1 billion pledge tied to curtailing deforestation in the country. "I reckon they got sick of being bombarded," he said, "so they started picking up the phone. Imagine, the Norwegian ambassador picks up the phone to the [Indonesian] president and says, 'Hey, we said we'd give you a billion dollars to stop doing this, and I turn on the BBC and the Tripa's up in flames. Is this a joke or what?' So the president picks up the phone to his carbon guy, and the carbon guy thinks, 'Thanks, folks, you've enabled me to do my job.' So he then goes and sends a team out, calls the minister of the environment, says, 'Hey, I don't have the capacity to prosecute people, but you do, so let's do this.' It gets things started."

In February of 2020, I opened my inbox to find an email from Emma Rae Lierley, forest communications manager for the Rainforest Action Network. After six years of draping banners, sweating inside of orangutan suits, and chanting at shareholders, she and her colleagues had just been informed by PepsiCo that the company intended to adopt a sweeping new palm oil policy. It vowed to ensure that none of the oil it sourced would be linked to deforestation, peatland destruction, or human-rights or labor abuses, and it emphasized that its terms would apply to its entire supply chain, even repeat-offender Indofood. In addition, PepsiCo was committing to play a more proactive role in the industry, including by addressing the ongoing threats to the global treasure that is Sumatra's Leuser Ecosystem. Lierley and her colleagues were making plans to celebrate the win, she said. And then they'd get back to work.

EPILOGUE
Post-Pandemic Palm

THIS ISN'T the epilogue I'd intended to write. But as we all know, the spring and summer of 2020 haven't been what any of us could have imagined they would be. Not only have we New Yorkers been locked down in our homes for five months thanks to the virus known as COVID-19, but the last few weeks have seen our streets pulsing with masked protesters wearing Black Lives Matter T-shirts and chanting "I can't breathe." In its own horrible but hopeful way, though, this dramatic moment in time, with its confluence of global pandemic and long-overdue racial reckoning, seems an appropriate coda for *Planet Palm*.

Certainly the crisis has brought starkly home just how closely our own health is linked to that of the planet, and how dangerous our continued destruction of the natural world has become. Though the origins of the virus that has so far taken some 700,000 lives worldwide remain unknown—reports that it emerged from a wet market in the Chinese city of Wuhan are unconfirmed—there's little question that the pathogen passed to humans from bats, likely through another species. Between 60 and 75 percent of today's emerging infectious diseases come from animals, and over the past few decades, the number of such animal-to-human, or "zoonotic," transmissions has skyrocketed. (Ebola, HIV, SARS, and MERS all came from animals.) A third of these new diseases can be linked directly to deforestation and agricultural intensification, most of it involving tropical rainfor-ests. Replacing these biodiversity hotspots with plantations doesn't just deprive orangutans and hornbills of their homes, in other words, it also sends virus-carrying wildlife like bats in search of new habitat, forcing them into closer contact with humans. "Our consumption drives pandemic risk," Peter Daszak, a disease ecologist and president of the New York–based EcoHealth Alliance, told me a few weeks into our city's lockdown.

We now devote half of the world's habitable land to agriculture, and yet one in nine people worldwide—some 850 million men, women, and children—remains hungry. One in three is overweight or obese. At the same time, the United Nations Intergovernmen-tal Panel on Climate Change has reported that global temperatures are on track to rise by at least 1.5° Celsius between 2030 and 2052, leading to extreme weather, destroyed ecosystems, water scarcity, reduced crop production, and mass migration, among other devas-tating impacts. Tropical deforestation alone is responsible for some 8 percent of the world's annual greenhouse gas emissions—more than those of the European Union. We are also living through a mass extinction, with our ever-dwindling biodiversity threatening the crucial services—including fresh water, pollination, and pest and disease control—that nature provides for free. While executives and shareholders of agribusiness giants like Cargill, Wilmar, Sinar Mas, and Socfin count their millions, oil-palm laborers across the globe toil under dangerous conditions and still don't make enough to feed their kids. Clearly, we are going about much of this in the wrong way.

But if the pandemic has shown us anything, it is that we're far more capable of change than we ever realized. In the face of this global emergency, governments and societies have mobilized quickly to undertake the kinds of systemic behavioral shifts that previously would have been unthinkable. Presidents and prime ministers have locked down entire countries, while citizens on multiple continents have (mostly) taken to voluntarily donning face masks, standing six feet apart, and relocating their professional lives to their living rooms. With everything once considered "normal" now gone out the window, we have a unique opportunity to rethink business as usual and chart the way toward a more sustainable future, one that doesn't just limit the emergence of pandemics but also promotes more equitable societies and heads off the worst that climate change has in store.

We can start by overhauling our food systems. Malnutrition is now the leading cause of ill health and deaths globally, and the COVID crisis is only making matters worse. It is well past time that we replace agricultural systems that favor calories over nutritional value and that make highly processed foods cheap and ubiquitous while healthy ones remain out of reach for so many. If we have any hope of feeding 9 billion people by 2050, we must stop planting land with commodities destined for animals' stomachs and fuel tanks and instead cultivate crops meant for humans, in particular nutritious options like healthy grains, fruits, vegetables, and legumes.

This also means slamming the brakes on industrial palm oil development. After all, the global industry has come to be what it is not as a result of a well-thought-out plan, and not because consumers decided that they couldn't live without more of this previously unknown substance. Rather, it is the culmination of a series of decisions made over decades, by national governments looking to alleviate poverty, by corporations and politicians looking to line their pockets, and by poor farmers looking to earn enough to feed their families. We know that some two-thirds of the palm oil produced globally ends up in food. And unlike the culturally prized and vitamin-rich artisanal oil consumed in Africa and Brazil, the industrial stuff makes its way into ultra-processed chips and baked goods, or into doughnuts and other fried fare. The obesity and diabetes that result from diets heavy in such oil-soaked junk shorten our life spans and tax our health systems. (They also make people more vulnerable to COVID-19, as do the respiratory conditions that result from the fires set to clear land for plantation development.) We simply do not need this much industrial palm oil in the global diet. What we *do* need, desperately, are intact tropical rainforests that can store carbon, house the wildlife whose forced migration leads to disease transmission and ecosystem collapse, and shelter the genetic richness that we'll require to confront the biological challenges of the future.

Key to avoiding the worst effects of climate change will be preserving our remaining peatlands. Indonesia's forests and peatlands alone store some 30 billion tons of carbon—more than three years of total global emissions from fossil fuels—and yet oil-palm plantations there and in neighboring Malaysia already have replaced 7.5 million acres of this critically important terrain. The governments of countries in Southeast Asia and the Congo Basin in particular must place moratoriums on new permits for development on primary forests and peatlands and must enforce them vigorously. Concession maps and ownership information in these and other countries should be made available for the public to see. Such disclosures could help curtail the backroom deals and political corruption that too often come at the cost of local communities. Companies responsible for setting fires on such lands (or for paying others to do so) should be punished with steep fines.

Protecting land of high carbon and conservation value will mean intensifying efforts on acreage already planted, including through breeding programs that focus on increasing yields. Where they haven't already done so, plantation companies should introduce cover crops and integrated pest management systems to cut down on chemical use and preserve soil and water resources. Smallholder farmers who rely on oil palm for income should be provided with fertilizers and quality planting materials in order to increase their own yields, and they should be trained in agroforestry systems like those common to the oil-palm regions of Africa and Bahia. Their access to credit should be conditioned on the adoption of sustainable growing practices. Ultimately, these farmers should be helped to transition away from oil-palm cultivation and toward more climate- and nutrition-sensitive agricultural systems.

Achieving the world's climate goals will require tapping into the expertise and holistic sensibilities of our indigenous peoples, who, despite inhabiting just 22 percent of the planet's land, protect 80 percent of its biodiversity. These communities, which have lived harmoniously within our tropical rainforests and other wild places for generations, must be given land tenure rights and ensured representation in environmental decision-making, including in any plans to transfer land to agribusiness corporations. Mapping forests and introducing property-registration systems would enable governments to better track and identify those responsible for clearing and to evaluate landowner compliance with conservation requirements. In the Amazon, a program initiated in 2009 saw as many as 40 percent of indigenous farmers granted title to large areas, a move that helped to dramatically limit deforestation in the region. Brazil's environment ministry helped pair indigenous communities engaged in sustainable economic activities with responsible partners in the cosmetics, ecotourism, and agroforestry industries. Similarly, indigenous communities in Malaysian Borneo have been working to help inventory the species richness of their rainforests, creating a library of tens of thousands of plants and microbes, genetic resources that have been protected under the Nagoya Protocol on Access and Benefit Sharing, thus ensuring that the communities profit from any eventual applications.

Of course, we can't expect the countries that are home to the world's tropical rainforests and peatlands to sacrifice their economies for the rest of us. More governments could follow the lead of Norway, which in recent years has also provided financing to Brazil and other countries in exchange for their having successfully curbed deforestation. Governments also could integrate tropical deforestation concerns into national energy considerations, as the EU recently did with its biofuels policy. Modern technology, including satellites, drones, and GPS systems, mean that it's no longer feasible for companies or others to plead ignorance when it comes to problematic supply chains. An open-access platform called Trase, launched in 2016 by the Stockholm Environment Institute and the British NGO Global Canopy, tracks the production and trade of commodities, including palm oil, that are linked to tropical rainforests, enabling companies, governments, and individuals to identify potential environmental or social risks along the way.

In September of 2019, as fires raged in the Amazon, 230 institutional investors with more than $16 trillion in assets under management called on companies to implement anti-deforestation policies for all of their supply chains, and to set up monitoring systems and report annual progress. Other financiers and investors should likewise emphasize the environmental risks in their portfolios, engaging with companies to eliminate commodity-driven deforestation and dropping those that fail to do so. Norway's Government Pension Fund Global, for example, has now divested from more than sixty companies associated with deforestation, thirty-three of them involved in palm oil. BlackRock's Larry Fink should follow through on a recent pledge to put climate change at the center of the firm's investment strategies. With assets under management worth some $7 trillion, BlackRock could serve as a model to other investors in putting concerns about the environment at the forefront of their decision-making.

Consumer-goods companies must adopt strict no-deforestation policies that extend to both direct and indirect suppliers and that include time-bound commitments and implementation plans. They should chart their progress online and put in place grievance mechanisms so that third parties can report abuses and track any steps being taken to address them. Multinationals like Unilever and PepsiCo could use their buying power to demand that suppliers fulfill traceability requirements that go beyond the still relatively weak principles and criteria of the RSPO. Putting a price premium on deforestation-free products and working with local partners could help incentivize smallholders to adopt more sustainable practices.

Finally, consumers concerned about the environmental and social fallout of palm oil should raise their voices to demand more transparency from the industry. Shareholders made aware of reputational, legal, and other risks in their companies—whether via protests, social media campaigns, or otherwise—tend quickly to get serious about ensuring that reforms are made. (Executives at Malaysia's Sime Darby, for example, screened the Emma Thompson–narrated orangutan video at their 2019 shareholder meeting, a clear indication of their sensitivity to consumer engagement.) Bold moves like Iceland's withdrawal from the sector also can send a signal to the industry that it had better clean up its act or face challenges accessing global markets.

• • •

In the end, it may turn out that some solutions to the many problems posed by the modern palm oil industry will emerge thousands of miles from the tropical rainforests of Southeast Asia or Africa. Six weeks before the onset of COVID-19 confined all of us to our homes, I boarded a flight to Madison, Wisconsin, touching down on a snowy February night. The next morning, I made my way to a laboratory located in a research park near the city's sprawling university campus, where two twinkly-eyed PhDs walked me through their plans for replacing the controversial commodity.

Tom Kelleher and Tom Jeffries earned their PhDs at Rutgers University, in New Jersey, in the 1970s, and met soon after at a gathering of microbiologists. Jeffries, a former national president of the Society for Industrial Microbiology and Biotechnology, went on to enjoy a long career at the U.S. departments of energy and agriculture, where he worked on the development of cellulosic fuels. Kelleher, a Cambridge native with a thick Boston accent, spent decades in the biopharmaceuticals industry, retiring from Amgen in 2014. He was the lead inventor of the manufacturing process for the antibiotic daptomycin and oversaw operations at two Italian factories dedicated to production of the drug, now known by its brand name, Cubicin.

A few years ago, Kelleher signed on as CEO of Xylome, the company that Jeffries founded in 2007 with a plan to manipulate the genetic pathways of different yeasts so as to create sustainable products. Having sequenced the genome of his preferred "bug," *Lipomyces starkeyi*, Jeffries worked with his team over three years to duplicate and rearrange its genes, finishing with a yeast that is capable of producing copious amounts of oil. Today, Xylome uses a fermentation process similar to that for brewing beer, with the engineered yeast metabolizing raw materials to produce oils with chemical profiles nearly

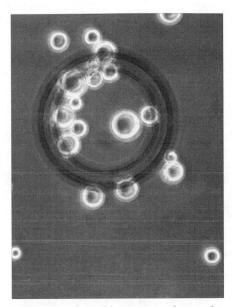

Synthetic palm oil being manufactured inside the yeast known as *Lipomyces starkeyi.*

identical to those of palm oil and palm kernel oil. The company has patented its effective strains and filed to be certified GRAS, or Generally Regarded as Safe, by the U.S. Food and Drug Administration. Peering into a microscope in their lab, I looked down on an assemblage of golden-hued blobs— palm oil in the making—and later rubbed some of the waxy, odorless end product into the back of my hand, watching it dissolve like a store-bought cream from a jar.

In 2015, Xylome was awarded a Small Business Innovation Research grant from the National Science Foundation for its proposal to create a palm oil equivalent for use in biodiesel using waste generated by ethanol plants. More recently, the company had received a three-year grant from the Department of Energy's National Renewable Energy Lab, for which it aims to develop an energy-dense biodiesel using corn stover (the cobs, husks, leaves, and stalks left in the field after the harvest) and cellulose as a feedstock. Not only is stover more efficient at converting energy from the sun than is the cornstarch normally used for ethanol, it has the advantage of not also being a food source. Jeffries and Kelleher have been collaborating with ethanol producers, among them a biofuels giant called POET, which is based in Sioux Falls, South Dakota, on a plan to use the waste generated by ethanol plants as a nutrient source for Xylome's oil-producing yeast.

Their ultimate vision is for a bioreactor to be sited at every ethanol plant in the country, cooking up net-carbon-neutral fuel. The trucks that today haul in corn for processing into ethanol could also deliver the stover needed to nourish their yeast, and the rail cars that cart the ethanol away could increasingly transport palm oil biodiesel.

The company is among a handful of groups worldwide that are working on initiatives to replace plantation-grown palm oil with synthetic versions. Breakthrough Energy Ventures, the multibillion-dollar fund established by Bill Gates and others to support innovations that combat climate change, recently invested $20 million in C16 Biosciences, a New York City–based startup with a plan similar to Xylome's for brewing synthetic oil, and researchers at the University of Bath, in England, hope to produce palm oil from the yeast *Metschnikowia pulcherrima*, using plant biomass broken down by microwaves as a feedstock.

Tom Kelleher's daughter, Nicole, is also on a mission to replace palm oil. A medical doctor, mother of two, and breast cancer survivor, the thirty-eight-year-old recently launched a company called Sarnaya with an aim toward making beauty products that are safe for humans and the planet. She had calculated that by replacing all of the palm oil currently used by the world's hair- and skincare makers with a synthetic, yeast-manufactured version, you could spare some 14 million tons of carbon emissions annually, the equivalent of eliminating 4 million around-the-world flights. Kelleher already was in talks with beauty and personal-care companies, some of which were experimenting with samples of her oil and with whom she expected to sign supplier agreements soon.

Jeffries and Kelleher senior, meanwhile, were working to get the price of their food-grade "biosimilar" below that of the plantation-grown variety, and had been in discussions with Colgate-Palmolive,

Procter & Gamble, Unilever, PepsiCo, and General Mills, among others, about potential partnerships. Though the pair had completed successful trial fermentations at a contract manufacturer in Milwaukee, their hope was to put together a consortium of three to five major palm oil buyers in order to justify the construction of a 100-ton stainless-steel bioreactor. They would use the same engineering specs to build additional plants as demand grows.

When it comes to addressing the myriad challenges facing the planet, the eternal optimists at Xylome told me they were just getting started. "The bio-concept of the future is that you don't just make something and throw all the waste away," said Kelleher, who added that the company hopes to use the debris from the palm oil fermentations to grow a fish feed that could help spare our fast-depleting rivers and oceans. "You take all the waste and make a new product, you take all the waste and make a new product," he said. "That's the way natural ecosystems work: all the waste products are used by some other organism."

A rethinking of the capitalist system that generates all that waste may also be in the offing. In early June, as Black Lives Matter protests convulsed cities across the United States, activists in the British city of Bristol fixed ropes to the statue of a slave trader named Edward Colston and proceeded to topple him from his plinth. Colston, who was born in Bristol in 1636, had worked for the Royal African Company, which for decades held the English monopoly on the trafficking of humans, eventually serving as its deputy governor. The cheering mob rolled his likeness through the Bristol streets and to the harbor's edge, giving it a final dramatic shove into the River Avon. Within a few days, protesters at Oxford University had launched a campaign calling for the removal of a statue of Cecil Rhodes; a 150-year-old statue of King Leopold II had been taken away from a public square in Antwerp; and another, in the city of Ghent, had been doused in red paint. Activists in Brussels had climbed atop a Leopold statue there and, while chanting "murderer" and "reparations," mounted a giant flag of the Democratic Republic of Congo. By June 12, calls were being made for the renaming of Leverhulme Park, a sports and recreation center in Bolton, on account of its namesake's own links to forced labor.

It's unclear just what fate England's statues and other memorials to William Lever will eventually meet, but in the meantime a remediation effort is well under way at the original site of the soap-maker's most egregious crimes. In 2012, a Congolese environmentalist named René Ngongo, formerly the head of that country's Greenpeace office, teamed with a Dutch artist named Renzo Martens to found an artists' cooperative with the express purpose of redressing the economic and cultural harms inflicted by Lever so long ago.

To get to Lusanga, the first of the industrialist's five original circles (and the one he'd quickly rechristened Leverville), I had to catch a flight to Kinshasa and then board a nineteen-seater for a journey over dense forest, finally landing in a buzzy little town called Kikwit, some four hundred miles to the southeast. I climbed out of the craft into the blinding sunshine and made my way across a sandy airfield and into a dark office, where the handwritten tickets, broken-down furniture, and casual demands for cash bribes aptly previewed the lessons that Lusanga has to teach about extractive economies and the people they leave behind. A two-hour drive along a rutted track brought me to a clearing dotted with huts of mud and thatch, plus a handful of mud-brick homes that had been built by Lever's men. This particular site, which most recently operated under the auspices of the Plantations Lever au Zaire (PLZ, formerly PLC, for Plantations Lever au Congo), had been sold by Unilever in 1990 and ceased operations a decade later.

I settled into a rustic bamboo hut—for all of Lever's boasting about state-of-the-art workers' camps, Lusanga, more than a century later, offers neither electricity nor indoor plumbing—and walked toward an open-air atelier, where twenty or so Congolese were bent over mounds of river clay, pinching and massaging it into expressive eyes and lips. Birdsong floated into the two-story structure, which was set alongside the gently coursing waters of the Kwilu River. (It was past these shores that Lever had chugged back in 1924, noting contentedly the "chubby piccaninnies all happy and smiling.") With the help of a few artists from Kinshasa, Ngongo and Martens had recruited a dozen members of what would become the Congolese Plantation Workers' Art League, or CATPC, based on their performance at an open audition. Ranging in age from twenty-three to eighty-seven, the artists are mostly poor farmers and oil-palm cutters. Many are the children and grandchildren of men who once toiled for Lever.

Martens' involvement in the program is an extension of a project he began more than a decade ago, when he spent three years filming a documentary of sorts, called *Episode 3: Enjoy Poverty*, in settings across the DRC. Released in 2008 to considerable acclaim, the provocation features footage of bloviating World Bank executives and obfuscating plantation managers, as well as interactions with impoverished laborers, forcing viewers to face uncomfortable truths about issues such as colonialism, racism, development aid, and the global capitalism system. Through a related project called the Institute for Human Activities, Martens and the artists of the CATPC established the Lusanga International Research Centre for Art and Economic Inequality, comprising the atelier and a spare conference hall set above it. Since 2017, the center has also served as improbable home to a modernist art museum—an iconic "white cube"—designed by OMA, the renowned Rotterdam firm founded by Rem Koolhaas. By putting an art museum in Lusanga, Ngongo and Martens explained, the group intends to upend the tradition of wealth and cultural capital being extracted from plantations for the benefit of people on the other side of the world. "Plantations like these have funded the very existence of the white cube," said Martens, who likes to point out that Unilever itself long underwrote a series of high-profile exhibits at London's Tate Modern, commissioning artworks by the likes of Bruce Nauman and Ai Weiwei for display in its cavernous Turbine Hall.

Once the artists have crafted their sculptures from river mud, the pieces get digitally scanned, and the resulting data is sent to Amsterdam. There, technicians employ 3D printers to create molds of the works, which are cast in unsweetened chocolate made from African cacao beans. Like palm oil, cacao has a long history of being produced on stolen land and relying on abusive labor practices. The sculptures, which have been exhibited in galleries and museums from Berlin and Amsterdam to Tokyo and New York (the *Times* named a 2017 show at the SculptureCenter, in Queens, among the best art of the year), have so far generated more than $100,000. All of the money goes back to the cooperative, which has used it to buy land and seedlings. The community is in the process of replacing the decades-old oil-palm monoculture with an integrated farm, or "post-plantation," intended to produce food for local families rather than a commodity destined for export. Their hope is to eventually secure the titles to some five thousand fertile acres.

"We are convinced that this model of eco-development," Ngongo told me, "financed by works of art made by Congolese artists, former workers in the Unilever plantations, can be reproduced in plantations elsewhere and so bring to a halt the paradox of the existence of indescribable poverty in zones that produce wealth for those in remote lands." (The inauguration of the museum, a ceremony that the community refers to as "The Repatriation of the White Cube," included a Skype call with members of Serbundo, the Indonesian plantation workers' union. The CATPC hopes to eventually connect with labor unions from across the world's plantation belt.)

The work itself speaks powerfully to Lever's enduring legacy. Irene Kanga, the twenty-five-year-old stepdaughter of a former PLZ employee, took as inspiration for her first sculpture the mass rape of the local Pende women, prompting the 1931 revolt described in Chapter 3. A dreadlocked thirty-three-year-old named Cedrick Tamasala, whose grandfather was orphaned after his father fell from a tree while harvesting fruit for Lever, named his original piece "How My Grandfather Survived." It features two figures, a larger, fatherly presence in robes holding open one side of a book, and a boy in shorts supporting the other side. Peeking up through the elder's hair is a pair of tiny horns. "The priest saved my grandfather," Tamasala told me, "but his mission was to destroy the culture. He was here to facilitate colonialism."

As the summer of 2020 lurches on, with statues connected to racial and economic injustice continuing to fall in cities around the world, many of us are realizing, perhaps for the first time, just how loudly history's drumbeats reverberate—how, for example, the damage so blithely wrought by the George Goldies and William Levers of the world can trickle down from one generation to the next. Just as many are calling for economic reparations in the United States, so too are palm oil communities across Africa, Southeast Asia, and Latin America demanding that the development banks and others

A sculpture made in the Democratic Republic of Congo references the country's links to an oppressive palm oil industry.

who have long underwritten their own disenfranchisement be finally made to pay a price. At the same time, the pandemic reminds us with each slowly passing month just how colossally we've screwed up our relationship with the natural world. Long-held notions about endless growth have come crashing into the reality of a finite, and deeply stressed-out, planet. As we consider how to move forward from this uniquely perilous place, we might look to the ones who've been quietly stewarding our rivers and forests all along, making space for the voices of indigenous peoples like the Brassmen, the Pende, and the Orang Rimba—voices to which we've failed to listen for so long.

ACKNOWLEDGMENTS

OVER THE too-many years that it's taken me to write this book, I've had help from dozens of people on multiple continents. I am particularly grateful to the farmers and plantation workers in villages around the world who took the time to tell me about their lives. I'm only sorry I couldn't fit all of their stories here.

Beginning with that original trip, my thanks to Devlin Kuyek, at GRAIN, for providing background on land grabs and steering me in the direction of Liberia. Of enormous help on the ground were researchers Ashoka Mukpo and Daniel Bucurus Krakue, of the Sustainable Development Institute. Marco Di Lauro provided breathtaking photographs and laughter along the way. I am grateful to Alfred Brownell, to government ministers Blamoh Nelson and Lewis Brown, and to county superintendents Milton Teajay and Alan Gbowee for taking the time to speak with me at length. And I owe a special debt to Cori Thomas and James Emmanuel "Kona" Roberts, for schooling me in Liberian history and politics and helping me with contacts in Monrovia.

Anyone writing about palm oil ends up spending a lot of time in Indonesia and Malaysia. I have many people to thank in those two countries. On Sumatra, Rudi Putra awakened me to the magic of the tropical rainforest, and Ian Singleton and Jen Draiss brought me into their extraordinary orangutan world. Farwiza Farhan provided insight into Acehnese politics, and Paul Hilton captured the glorious biodiversity with his camera. Thanks, too, to Kemal Jufri for moving images taken on a subsequent trip to the region. The investigators at Eyes on the Forest let me into their undercover existence and answered endless questions over days spent together on the road. A big thank you, with apologies, to Elok Sandhi, Herwin Nasution, and Lambok Simbolon, of the labor-rights organization OPPUK, and to Daryll Delgado, of Verité Southeast Asia, for the education they provided and for too many aborted trips. My Jambi-based translator knows how grateful I am to him. Revan and Irsan Pragustiawan, Hassan Basri, Abdullah Sani, and others from the Batin Sembilan and Orang Rimba communities graciously welcomed this outsider in and schooled me in their indigenous cultures. Thank you also to Norman Jiwan, Rukaiyah Rafiq, Feri Irawan, Yokyok "Yoki" Hadiprakarsa, Dean Ismail, and Shayne McGrath for guidance on the ground and on subjects including human rights, birdlife, agronomy, village politics, and more.

In Brazil, Jon Lewis provided an insider's view of Bahian culture and unmatchable companionship. Thank you to Ayrson Heráclito and to Mãe Bárbara and the residents of her coastal community for an immersive education in *candomblé*. Alicio Charoth came through with thoughtful conversation and an unforgettable meal. Thank you to Shirley Stolze for the beautiful photos.

Annie Bird was an invaluable source of information on Guatemalan history and politics. In Guatemala City, Laura Hurtado took several hours out of her day to discuss the impacts of the industry on local livelihoods and health, and helped link me to farmers in the Petén. Thanks to my translator and fixer Jeff Abbott, and to Guatemalans Saul Paau, Juan Manuel De la Cruz, Lorenzo Pérez Mendoza, Remigio Caal, Hermalindo Asigno, Karla Hernandez, and Maria Margarita Ivanez for spending time with me.

In Honduras, Walter Banegas, Yarleni Ortéz Mejía, and Marcelino Flores graciously invited me into their homes and spoke candidly about their lives. Other workers and former workers of Grupo Jaremar were generous with their time, and labor organizer Ahraxa Mayorga somehow never lost his patience despite my never-ending queries both during and after my visit. Thanks also to my sharp translator Oscar Orlando Hendrix Escalante and to ace local journalist Lourdes Ramirez for context and companionship.

In India, Dr. Anoop Misra welcomed me into his office and spent more time than he had discussing dietary concerns in the country. Thanks also to his associates Shubhra Atrey and Amrita Ghosh. I am grateful to Shravya Reddy for the introduction to her father, K. Srinath Reddy, who in turn steered me toward Suparna Ghosh Jerath, Shweta Khandelwal, and Suma Sajan at the Public Health Foundation of India. Conversations with Pawan Agarwal, of the Food Safety and Standards Authority of India; Kamal Kapoor, of KP Agro Oils; shop owner Sanjay Kumar; journalist Adarsh Garg; Cargill India's Siraj Chaudhry; and Subodh Jindal, of the All India Food Processors' Association, helped me to understand the complicated factors at play in the country. My knowledge about diet and nutrition in the Mexican context comes thanks largely to Alejandro Calvillo, of El Poder del Consumidor, and Luis Manuel Encarnación, at Contra-PESO Coalition. Saskia Heijnen provided expertise on human and planetary health from her office at the Wellcome Trust, in London.

Reaching a remote village in the Democratic Republic of Congo takes some doing. I am grateful to Nicholas Jolly, Laurens Otto, and Janke Brands, at the Amsterdam-based Institute for Human Activities, for help with logistics. In Lusanga, the artists of the Cercle d'Art des Travailleurs de Plantation Congolaise, in particular Cedric Tamasala, Matthieu Kiambu, Mbuku Kimpala, and Irene Kanga, took time away from their sculpting to tell me about their work. Thanks to René Ngongo, Renzo Martens, and Eléonor Hellio for conversations on art and economy by the river.

I couldn't have tracked the comings and goings of this far-flung industry without the help of many talented journalists working in the countries I've written about here. Thank you in particular to Wade Williams and Rodney Sieh in Liberia; Sunday Orji in Nigeria; Clare Rewcastle Brown, writing about Malaysia from the U.K.; Margie Mason and Robin McDowell, of the Associated Press, in Indonesia; A. Ananthalaksmi and Emily Chow, of Reuters, based in Malaysia and China, respectively; and Tom Johnson and his colleagues at the Gecko Project. Rhett Butler and the folks at Mongabay are an invaluable resource for anyone interested in rainforests. I am grateful to reporters Hans Nicholas Jong, Loren Bell, and John C. Cannon for their nuanced work, and to Rhett himself for always responding quickly and graciously, and for the photos I've used here.

Many in academia took time out of their busy schedules to share their expertise, among them, Joshua Linder and Case Watkins of James Madison University (I owe a particular debt to Case for his brilliant scholarship and the use of his map); Bhavani Shankar at the University of London's School of Oriental and African Studies; Richard Smith at London School of Hygiene & Tropical Medicine; Walter C. Willett, Frank B. Hu, and Qi Sun at Harvard University; Shauna Downs at Rutgers University; and Barry Popkin of the University of North Carolina. I am especially grateful to Wally Falcon, Roz Naylor, and Derek Byerlee, for their invaluable book *The Tropical Oil Crop Revolution*, and for the conversations that followed my reading of it. Derek Yach, formerly of the World Health Organization, spoke with me about his work on several occasions. Thanks also to

Peter Daszak, of the EcoHealth Alliance, and to Timothy D. Searchinger and Chris Malins for their work on biofuels.

This book wouldn't have been possible without the help of my many acquaintances in the NGO world. For assistance with contacts in various countries and for patiently answering my emails over the course of many months, I am grateful to Laurel Sutherlin, Gemma Tillack, Chelsea Matthews, Emma Rae Lierley, Robin Averbeck, and the other hard-working folks at the Rainforest Action Network; to Rick Jacobsen, Billy Kyte, Lela Stanley, and Jonathan Gant at Global Witness; to Jeff Conant and Andrew Fandino at Friends of the Earth US; to Doug Hertzler at ActionAid USA; to Glenn Hurowitz at Mighty Earth; to Sol Gosetti, Victoria Henry, and Maya Marewu at Greenpeace; to Annette Lanjouw at the Arcus Foundation; to Nigel Sizer and Brittany Wienke at the Rainforest Alliance; to Marcus Colchester and Patrick Anderson at the Forest Peoples Programme; to Sarah Fogel and Sarah Forrest at WWF; and to Gabby Rosazza and Eric Gottwald at Global Labor Justice (formerly the International Labor Rights Forum). Thanks also to Paul Pavol and Nasako Besingi, in Papua New Guinea and Cameroon, respectively, for bringing to my attention the situations in their countries.

Many thanks to the executives and others involved in the palm oil industry, for explaining how things are done at their companies and for answering my not-always-welcome questions. Among those who gave their time were Alexandra Palt, Rachel Barre, and Adélaide Colin of L'Oréal; Gay Timmons of Oh, Oh Organic; Boris Oak of EVOLVh; Perpetua George, Yeap Su Jeen, and Iris Chan Suiet Yeng of Wilmar International; Neil Blomquist, Prescott Bergh, Hans van den Heuvel, and Lia Huber of Natural Habitats; Anita Neville of Golden Agri-Resources; Simon Lord, Jamie Graham, and

Carl Dagenhart of Sime Darby; Viggy Ponnudurai and Matt Karinen of Golden Veroleum Liberia; Juan Marco Alvarez of REPSA; Sonia Mejia of Grupo Jaremar; Roger Pineda of Dinant; Chad Risley of Berg+Schmidt; Tom Kelleher and Tom Jeffries of Xylome; and Nicole Kelleher of Sarnaya. Thanks also to Unilever archivists Victoria Howard and Nicole Hubberstey, and to the RSPO's Stefano Savi, Danielle Morley, Darrell Webber, and Dan Strechay.

Parts of this book first appeared in magazines, and I am indebted to the editors who assigned those stories and helped finance my travels. Among them are Doug Barasch, George Black, Scott Dodd, and Janet Gold at *OnEarth*; Rene Ebersol, Jenny Bogo, and Mark Jannot at *Audubon*; Larry Kanter and Greg Emmanuel at *Men's Journal*; Celia Ellenberg at *Vogue*; Alex Testere and Stacy Adimando at *Saveur*; Roane Carey and Lizzy Ratner at *The Nation*; Eric Lach at newyorker.com; and Mike Hudson at the International Consortium of Investigative Journalists. Sam Fromartz, Tom Laskawy, and Brent Cunningham of the Food and Environment Reporting Network, provided invaluable editorial guidance and financial support on more occasions than I can count.

In 2017, I was fortunate enough to receive a generous fellowship from the Alicia Patterson Foundation. My thanks to Margaret Engel and the others at the foundation, without whom this book wouldn't have been possible.

Thank you to my brilliant fact-checker, Ben Phelan, for tracking down every date and detail, and to New Press production editor Emily Albarillo for her work ethic, her eagle eye, and her boundless patience. If errors remain, they are my own. I was lucky to have Sofia Perez, Marisa Robertson-Textor, and Cintia Chamecki available in New York for translating interviews, reports, and communications variously from Spanish, Russian, and Portuguese.

Marc Favreau, executive editor of The New Press, took this project on with enthusiasm and provided encouragement and patience—not to mention editing expertise—over many months. Thanks also to Brian Ulicky, Emily Janakiram, and Jay Pabarue for help with marketing and publicity, and to Michael Dwyer, Farhaana Arefin, and Kathleen May at Hurst Publishers, for their work on the U.K. edition of this book.

My agent, Sarah Lazin, saw something in this idea before anybody else did, and she displayed the patience of a saint over the many months it took me to produce a proposal and finally a book. She calmly kept me on track with warmth and humor. Thanks also to her assistants Catharine Strong and Margaret Shultz.

I'm grateful to my dad, who, before his death in 2014, constantly nagged me about writing a book, and to my incredible eighty-two-year-old mom, for understanding that motherhood doesn't cancel out other ambitions. She gamely drove from New Jersey to Brooklyn to watch my girls (and my dog) whenever I set off on another trip, and the series of emoji-filled texts that she invariably sent in the days that followed assured me that all was well back home. "What an example for your girls," she would write.

To those angels themselves, Daisy and Ella, for letting me go, over and over, and for becoming the deeply kind and probing young women they are today.

Finally, thanks to my husband, Bill, for juggling life back home, for believing in me (or pretending to) on the many occasions I didn't believe in myself, and for the volumes of laughter and love over these twenty-some years.

NOTES

Epigraph

vii **"For a colonized people"**: Frantz Fanon, *The Wretched of the Earth* (1961; reprint, New York: Grove Press, 2004), 9.

Prologue: Oil Crisis

1 **I was rattling**: I traveled to Liberia from February 21 to March 3, 2013.

1 **founded by freed American slaves:** Maybe not as high-minded as it sounds. Beginning in 1816—fifty years before the U.S. abolished slavery—the American Colonization Society sought to create a colony in Africa to which it could send slaves. It secured land in West Africa and began shipping slaves to the colony, which became the nation of Liberia in 1847. John Hanson Thomas McPherson, *History of Liberia* (Baltimore, MD: The Johns Hopkins Press, 1891).

2 **The phenomenon:** Fred Pearce, *The Land Grabbers: The New Fight Over Who Owns the Earth* (Boston: Beacon Press, 2012).

2 **I'd long admired:** The more I learned about her, the less I found to admire. See Jocelyn C. Zuckerman, "Lipstick's Steep Price: Africa's Vanishing Forests," *Salon.com*, December 11, 2013.

2 **having spent twelve years:** I worked at *Gourmet* from 1996 to 2008.

2 **In one village:** The village was called Pluoh. I visited several villages impacted by the palm oil companies Golden Veroleum Liberia (GVL) and Sime Darby during my stay. GVL is a subsidiary of the Verdant LP Fund, whose lead investor is Singapore-based Golden Agri-Resources, or GAR. Sime Darby is based in Kuala Lumpur.

3 **A Monrovia-based lawyer:** I interviewed Alfred Brownell, of Green Advocates, in his Monrovia office on February 27, 2013. We have since corresponded frequently via phone and email.

3 **It had signed:** This information is based on the August 2010 concession agreement between the government of Liberia and GVL, and was confirmed by Matt Karinen, director of GVL at the time. It specifies 350,000 hectares, which is equal to 864,868 acres.

3 **In my twenties:** I served as a Peace Corps volunteer in Kenya's Busia district from September 1991 to November 1993.

5 **Sinoe County, though home:** According to the "2008 Population and Housing Census, Final Results," put out by the Liberia Institute of Statistics and Geo-Information Services.

5 **taking some 250,000 lives:** According to BBC News World Africa.

5 **The Ebola outbreak:** Liberia's Ebola outbreak peaked in August and September of 2014. Ashoka Mukpo, the researcher who fell ill, was working as a freelance journalist for NBC at the time.

5 **I know well that:** "Agriculture and Food: Overview," The World Bank. https://www.worldbank.org/en/topic/agriculture/overview.

5 **a point echoed by:** I spoke with executives from both Sime Darby and GVL, as well as with local and national politicians, while in Liberia and afterward.

6 **with roughly half of all products:** According to the World Wildlife Fund. https://www.worldwildlife.org/pages/which-everyday-products-contain -palm-oil.

6 **Palm oil alone:** Specifically, 35 percent. *OECD-FAO Agricultural Outlook, 2019–2028*, Chapter Four: "Oilseeds and Oilseed Products," 143.

6 **Ditto the moisturizer:** The generic versions of these products tend to include palm oil derivatives, including oleochemicals and glycerol. Based on interviews conducted with executives and lab technicians at L'Oréal headquarters in Paris on October 13 and 14, 2016. See also Hillary Rosner, "Palm Oil Is Unavoidable. Can It Be Sustainable?" *National Geographic*, December 2018. For information about palm oil–free versions of these products, see websites such as Bustle.com, Beautycalypse.com, and Selvabeat.com.

6 **in the non-dairy creamer:** According to the Malaysian Palm Oil Board. http://palmoilis.mpob.gov.my/publications/TOT/tt196.pdf. See also, Coffee-Mate label. https://www.coffeemate.com/products/powder/french-vanilla.

6 **in the doughnuts:** According to various doughnut labels, including Entenmann's. See also, Dunkin' Donuts Palm Oil Sourcing Guidelines, at https://www.dunkinbrands.com.

6 **in the baby formula:** Fatty acids from palm oil are a common ingredient in infant formula. See, for example: John B. Lasekan, Deborah S. Hustead, Marc Masor, and Robert Murray, "Impact of palm olein in infant formulas on stool consistency and frequency: a meta-analysis of randomized clinical trials," *Food & Nutrition Research* 61 (June 2017).

6 **in the dog food:** According to the Malaysian Palm Oil Board. http://palmoilis.mpob.gov.my/V4/palm-oil-in-dog-food-itex-2015. See also, "Palm

Kernel Meal in UK Pet Food Contributing to Deforestation," Petfoodindustry
.com, May 2011.

6 **It's in the Nutella:** According to Nutella. https://www.nutella.com/us
/en/nutella-palm-oil.

6 **including in the feed:** Palm-kernel meal is a common ingredient in animal feeds. In addition, a type of fractionated palm oil often is fed to lactating
cows. I am grateful to Chad Risley, general manager/CEO at Berg+Schmidt
America, in Libertyville, Illinois, for walking me through the role of various
palm oil and palm-kernel oil derivatives in animal feeds.

6 **India, now the world's:** For imports and exports of palm oil and other
commodities, I have relied on IndexMundi. https://www.indexmundi.com.

6 **Worldwide, production:** For production figures, I have relied on the
United States Department of Agriculture Foreign Agricultural Service
(FAS). Production in 2005: 36 million metric tons. In 2020: 72.2 million met-
ric tons.

6 **oil-palm plantations now cover:** According to the Food and Agriculture Organization of the United Nations, or FAO.

6 **an area larger than:** 27 million hectares = 104,247 square miles. New
Zealand = 103,483 square miles.

6 **Indonesia and Malaysia, which together:** Indonesia = 43.5 million
metric tons. Malaysia = 19.3 million metric tons. 43.5 + 19.3 = 62.8 million.
0.85 of 72.2 million = 61.3 million metric tons.

6 **they're expanding to:** John C. Cannon, "Palm Oil Interest Surges in
Papua New Guinea," *Mongabay*, November 19, 2014; Rod Harbinson, "On a
Philippine Island, Indigenous Groups Take the Fight to Big Palm Oil," *Mong-
abay*, July 11, 2019; "PM Commends PG Palm Oil Operation," *Papua New
Guinea Post Courier*, March 2, 2020.

6 **and farther afield:** See Chapter 6 in this book.

7 **that's roughly twenty pounds:** 72 million metric tons = 159 billion
pounds. 159 billion pounds divided by 7.8 billion people = 20.4 pounds per person.

7 **"You're soaking in it":** According to Colgate Corporate History. http:
//www.colgate.com/app/Colgate/US/Corp/History/1961.cvsp.

7 **in the five decades since:** Figures according to IndexMundi: 2005
imports = 596,000 metric tons. 2019 imports = 1.565 million metric tons.

7 **thanks in part:** In 2015, the U.S. Food and Drug Administration
determined that partially hydrogenated oils (PHOs) were no longer "Generally Recognized as Safe," or GRAS. After June 18, 2018, manufacturers were
no longer allowed to add PHOs to foods. https://www.fda.gov/food/food

-additives-petitions/final-determination-regarding-partially-hydrogenated
-oils-removing-trans-fat.

7 **In addition to:** R.H.V. Corley and P.B. Tinker, *The Oil Palm*, 5th ed. (West Sussex, U.K.: John Wiley & Sons, 2016). I have relied on Corley and Tinker's incomparable scholarship throughout this book.

7 **one that took me:** In the course of reporting this book, I traveled to Africa (Liberia and the Democratic Republic of Congo); Europe (France, England, and Italy); Asia (Malaysia, Indonesia, India, and Thailand); and South and Central America (Brazil, Mexico, Ecuador, Guatemala, and Honduras).

8 **In the same way that:** Mark Kurlansky, *Salt: A World History* (New York: Penguin, 2002); Sven Beckert, *Empire of Cotton: A Global History* (New York: Alfred A. Knopf, 2014); Sidney W. Mintz, *Sweetness and Power: The Place of Sugar in Modern History* (New York: Penguin Random House, 1985).

8 **Today, palm oil stands:** Prof. Boyd A. Swinburn, MD, Vivica I. Kraak, PhD, Prof. Steven Allender, PhD, Vincent J. Atkins, Phillip I. Baker, PhD, Jessica R. Bogard, PhD, et al., "The Global Syndemic of Obesity, Undernutrition, and Climate Change: *The Lancet* Commission Report," *The Lancet*, January 27, 2019.

8 **A perennial plant:** Corley and Tinker, 1–2, 9; Derek Byerlee, Walter P. Falcon, and Rosamond L. Naylor, *The Tropical Oil Crop Revolution: Food, Feed, Fuel, & Forests* (Oxford, U.K.: Oxford University Press, 2017), 17.

8 **"They were like":** Graham Greene, *A Burnt-Out Case* (1960; reprint: New York: Penguin Books, 1977), 138.

8 **Archaeological findings suggest:** Corley and Tinker, 1.

8 **the Greek historian Herodotus:** "Egyptian Mummification," Spurlock Museum of World Cultures, University of Illinois.

10 **When the agriculturalists left:** Corley and Tinker, 2–3.

10 **It's for this reason:** Susan M. Martin, *Palm Oil and Protest: An Economic History of the Ngwa Region, South-Eastern Nigeria, 1800–1980* (Cambridge, U.K.: Cambridge University Press, 1988), 13.

10 **The plants, which begin:** Corley and Tinker, 103, 261.

10 **yielding considerably more:** Corley and Tinker, 10.

10 **In its unrefined form:** *Fats and Oils in Human Nutrition*, Chapter 14: "Non-glyceride constituents of fats" (Rome: FAO, 1994). http://www.fao.org/3/v4700e/V4700E00.htm.

10 **In parts of rural Liberia:** Kwasi Poku, *Small-Scale Palm Oil Processing in Africa* (Rome: FAO, 2002). See also, Jules Marchal, *Lord Leverhulme's Ghosts: Colonial Exploitation in the Congo* (London: Verso, 2008), 3.

10 **With the trees reaching:** Corley and Tinker, 309.

11 **slow the development:** Corley and Tinker, 460.

11 **After a final pass:** Tola Atinmo, PhD, and Aishat Taiwo Bakre, MSc, "Palm Fruit in Traditional African Food Culture," *Asia Pacific Journal of Clinical Nutrition*, December 2003. See also, Martin Lynn, *Commerce and Economic Change in West Africa: The Palm Oil Trade in the Nineteenth Century* (Cambridge, U.K.: Cambridge University Press, 1997), 46–48.

11 **Europeans sailing down:** Corley and Tinker, 3–4.

11 **"This palm oil is":** J. Barbot, "A Description of the Coast of North and South Guinea and of Ethiopia Inferior, Vulgarly Angola," in A. Churchill and J. Churchill (eds.), *A Collection of Voyages and Travels*, 6 vols. (London: 1732), vol. V, 204.

11 **The African American culinary:** Jessica Harris, *The Africa Cookbook: Tastes of a Continent* (New York: Simon & Schuster, 1998), 348.

11 **In his 1958 classic:** Chinua Achebe, *Things Fall Apart* (1958; reprint, New York: Anchor Books, 1994), 7.

11 **"Those whose palm-kernels":** Achebe, 26.

12 **By the 1890s:** Corley and Tinker, 4. See also, Chapter 1 in this book.

12 **a curious foreshadowing:** Toyin Falola, *Colonialism and Violence in Nigeria* (Bloomington: Indiana University Press, 2009), Chapter 1.

12 **Europeans originally sourced:** Corley and Tinker, 4.

12 **Tinplate makers came:** Makalé Faber Cullen, "The Oil Palm Kernel and the Tinned Can," *Limn*, May 2014.

12 **Eventually, tins made:** Faber Cullen.

12 **by the end of the nineteenth century:** Lynn, 3. See also, Frederick Pedler, *The Lion and the Unicorn In Africa: The United Africa Company 1787–1931* (London: Chaucer Press, 1974), 173.

12 **Whereas in 1870:** Marchal, ix.

12 **He followed in the footsteps:** See Chapters 1 and 3 in this book.

13 **En route, palm oil:** Lynn, 32.

13 **Escaped and freed slaves:** Judith A. Carney and Richard Nicholas Rosomoff, *In the Shadow of Slavery: Africa's Botanical Legacy in the Atlantic World* (Berkeley: University of California Press, 2009), 40, 85, 201, 254.

13 **The legacies—:** The companies Sime Darby, Socfin, and United Plantations Berhad were originally founded by these men.

13 **"The Indians who labored":** Tash Aw, *We, the Survivors* (New York: Farrar, Straus & Giroux, 2019), 24.

14 *Elaeis guineensis* **thrives:** Corley and Tinker, 3.

15 **Moving on from Honduras:** See Chapter 6 in this book.

15 **"People are using":** See Chapter 7 in this book.

15 **In 2015, an extended episode:** *The Cost of Fire: An Economic Analysis of Indonesia's 2015 Fire Crisis* (Jakarta: World Bank, 2016). http://pubdocs
.worldbank.org/en/643781465442350600/Indonesia-forest-fire-notes.pdf.

16 **A few weeks into the crisis:** Georgia McCafferty, "Indonesia Begins Evacuation of Babies from Haze-Affected Regions," *CNN*, October 1, 2015.

16 **Indonesia is home:** Corley and Tinker, 78.

16 **Much of the blame:** See Chapter 8 in this book.

16 **Malaysia's $16 billion industry:** According to Malaysia's Plantation Industries and Commodity Minister and the Malaysian Palm Oil Council (68 billion Malaysian ringgit = $16.28 billion).

16 **the European Union announced:** Reuters staff, "Indonesia Threatens to Quit Paris Climate Deal over Palm Oil," *Reuters*, March 27, 2019.

17 **the $65 billion business:** Research, Zion Market. "Global Report: Palm Oil Market Size & Share Estimated to Touch the Value of $92.84 Billion in 2021." *GlobeNewswire News Room*, "GlobeNewswire," July 30, 2019, www
.globenewswire.com/news-release/2019/07/30/1893425/0/en/Global-Report
-Palm-Oil-Market-Size-Share-Estimated-To-Touch-the-Value-Of-USD-92
-84-Billion-In-2021.html.

17 **It recently emerged:** See Chapter 9 in this book.

17 **"There's an awful lot":** I interviewed Sizer in his New York office on July 12, 2016.

18 **In 2019, hundreds:** *2019 Global Assessment Report on Biodiversity and Ecosystem Services*, coordinated by the Bonn-based Intergovernmental Science-Policy Platform on Biodiversity and Ecosystem Services (IPBES).

18 **Tropical rainforests:** E.O. Wilson and Frances M. Peter, eds, *Biodiversity*. Chapter 3: "Tropical Forests and their Species Going, Going . . . ?" by Norman Myers (Washington, DC: National Academies Press, 1988).

18 **some 2,500 square miles:** European Union's Atmosphere Observation Program, referenced by Yoga Rusmana, "Forest Fires from Indonesia Worse Than the Amazon, EU Says," *Bloomberg News*, November 26, 2019. Per Bloomberg, more than 4,000 kilometers, or 2,500 miles.

18 **"You'd see the buses":** Aw, 316–17.

Chapter 1: Goldie Goes In

23 **"The elders consulted":** Chinua Achebe, *Things Fall Apart* (1958; reprint, New York: Anchor Books, 1994), 138.

23 **In my mind's eye:** J.E. Flint, *Sir George Goldie and the Making of Nigeria* (Oxford, U.K.: Oxford University Press, 1960), 5. Flint's book, based on personal interviews, letters, and copious government documents, is the only comprehensive source on Goldie's life. I am reliant on him for much of the information in this chapter.

23 **The fact is:** Thomas Pakenham, *The Scramble for Africa: The White Man's Conquest of the Dark Continent from 1876 to 1912* (New York: Random House, 1991), 184. I recommend Pakenham's mind-bogglingly thorough and erudite book to anyone with even a passing interest in Africa. I have relied heavily on Pakenham for information about the events across the continent during the period covered in this chapter.

23 **"I was like":** Flint, 4.

24 **Then there were:** Flint, 5.

24 **surviving, like the rest:** Alistair Horne, *The Fall of Paris: The Siege and the Commune 1870–71* (New York: Penguin Books, 1965), 176–93. See also, Stéphane Hénaut and Jeni Mitchell, *A Bite-Sized History of France: Gastronomic Tales of Revolution, War, and Enlightenment* (New York: The New Press, 2018), 221–8. Though I don't have documentary evidence that Goldie and Elliot ate dog and rat, the likelihood is extremely high, given that little other protein was available in the city, and most residents were forced to resort to the two.

24 **Holland, Jacques & Company:** In addition to Flint, see K. Onwuka Dike, *Trade and Politics in the Niger Delta 1830–1885: An Introduction to the Economic and Political History of Nigeria* (Oxford, U.K.: Oxford University Press, 1956), from p. 208.

24 **In the fifteenth century:** Dike, 1–5; According to the Metropolitan Museum's Heilbrunn Timeline of Art History. https://www.metmuseum.org/toah.

24 **By 1792:** According to the Slave Voyages website, a collaborative project sponsored by the National Endowment for the Humanities, Emory Center for Digital Scholarship, the University of California at Irvine, the University of California at Santa Cruz, and the Hutchins Center of Harvard University. https://slavevoyages.org.

24 **Roughly half of them:** According to Slave Voyages and the Liverpool Museum's History of Slavery. https://www.liverpoolmuseums.org.uk/history-of-slavery/europe.

24 **Though Parliament outlawed:** Martin Meredith, *The Fortunes of Africa: A 5000-Year History of Wealth, Greed, and Endeavor* (London: Simon & Schuster, 2014), 218.

24 **Between 1750 and 1780:** The Abolition Project, East of England Broadband Network. http://abolition.e2bn.org/slavery_45.html. See also, David Richardson, "The Slave Trade, Sugar, and British Economic Growth, 1748–1776," *Journal of Interdisciplinary History* No. 4 (Spring 1987), The MIT Press, Volume 17.

25 **No one ever asks:** David Kaiza is based in Kampala, Uganda.

25 **The Niger Delta:** Dike, Chapter 1.

25 **Over the course of:** Martin Lynn, *Commerce and Economic Change in West Africa: The Palm Oil Trade in the Nineteenth Century* (Cambridge, U.K.: Cambridge University Press, 1997), 26.

25 **City-states such as Bonny:** Obaro Ikime, *Merchant Prince of the Niger Delta: The Rise and Fall of Nana Olomu, Last Governor of the Benin River* (New York: Africana Publishing Corporation, 1969), 5. See also, The Abolition Project, British Involvement in the Transatlantic Slave Trade, East of England Broadband Network. http://abolition.e2bn.org/slavery_45.html.

25 **"I believe":** Mary Kingsley, *Travels in West Africa* (1897; reprint, Washington, DC: National Geographic Society, 2002), 59.

26 **Europeans referred:** *The Naval and Military Magazine*, Vol. II, 1827, 313. See also, Dike, 10.

26 **For the duration:** Dike, 8.

26 **Upon his appointment:** Thomas Wright, *The Life of Sir Richard Burton* (Library of Alexandria, 1906), 176. Retrieved through Project Gutenberg.

26 **In the meantime:** Dike, 49.

27 **By 1850:** Flint, 10.

27 **At the same time:** Lynn, 28–9.

27 **By the time:** Lynn, 154.

27 **"I have seen it written":** A.C.G. Hastings, *The Voyage of the Day Spring* (London: John Lane the Bodley Head, 1926), 217, quoted in Dike, 112.

27 **Another John:** Dike, 50. See also, Lynn, 83.

27 **He is said:** Martin Lynn, "The Profitability of the Early Nineteenth-Century Palm Oil Trade," *African Economic History*, no. 20 (Madison: University of Wisconsin Press, 1992), 77–97.

27 **So potentially lucrative:** Henry Stanley, *Through the Dark Continent*, published by Stanley in 1899.

27 **the men chasing palm oil:** Pakenham, xxii.

28 **"palm-oil ruffians":** Dike, 60. See also, Lynn, *Commerce and Economic Change in West Africa*, 89.

28 **Edward Nicolls:** Lynn, *Commerce and Economic Change in West Africa*, 89.

29 **Eight-year-old Goldie:** Flint, 4.

29 **Looking on:** Lynn, *Commerce and Economic Change in West Africa*, 66, 92.

29 **unlike captives:** Ed Emeka Keazor, *Nigeria and the Royal Niger Company: 1879–1900* (Academia.edu, 2014). See also, Lynn, "The Profitability of the Early Nineteenth-Century Palm Oil Trade"; and Toyin Falola, *Colonialism and Violence in Nigeria* (Bloomington: Indiana University Press, 2009), 4–5; and Martin, 49.

29 **The U.K. traders:** Lynn, *Commerce and Economic Change in West Africa*, 84. See also, Dike, 49.

30 **Like the traders:** Dike, 102–6. See also, Lynn, *Commerce and Economic Change in West Africa*, 72–3.

30 **"The kings of the countries":** Cantor Lectures, "Solid and Liquid Illuminating Agents," *Journal of the Society of the Arts* vol. 31, 858–62. Retrieved through Google Books.

30 **Middlemen often were:** Dike, 34–42.

30 **Competition among:** Falola, 4. Also, Dike, 42.

30 **"They are a people":** Percy Amaury Talbot, *Tribes of the Niger Delta* (Charlottesville, VA: Sheldon Press, 1932), 9, quoted in Dike, 46.

31 **Things had begun to change:** All per Lynn, *Commerce and Economic Change in West Africa*, and Dike.

31 **Annoyed at losing out:** Flint, 21.

31 **In response:** Dike, 93, 207.

31 **As tensions rose:** Pakenham, 182; Dike, 203–9.

32 **In an 1871 memo:** Dike, 198.

32 **"Their property":** Dike, 198, quoting Foreign Office letter from December 3, 1871.

32 **"In all the rivers":** "A View of Old Calabar," *Chambers's Journal* 51, no. 2, 526, quoted in Flint, from Parliamentary Papers 1875. LXV, MS. p.3.

33 **"Many years ago":** Ockiah and other chiefs of Brass to Foreign Office, July 7, 1876, quoted in Flint, 28. See also, Pakenham, 195.

34 **"a convivial fellow":** Pakenham, 191.

34 **"The coast is pestilential":** Pakenham, 188.

34 **The NAC:** Pakenham, 186.

35 **"My dream as a child":** C.H. Currey, *The British Commonwealth Since 1815* (Sydney, 1951), 49, quoted in Dike, 208. See also, Kwasi Kwarteng, *Ghosts of Empire: Britain's Legacies in the Modern World*, chapter 14; and Washington A. J. Okumu, *The African Renaissance: History, Significance and Strategy* (Trenton, NJ: Africa World Press, 2002), 44. Retrieved through Google Books.

35 **a young prodigy:** Dike, 183. See also, Flint, 27.

35 **"one Jaja":** Dike, 184, from Foreign Office letter from Burton to Russell, August 8, 1864. See also, Lynn, *Commerce and Economic Change in West Africa*, 179–180.

35 **"Gentlemen":** Dike, 188, quoting Foreign Office letter from Jaja, September 14, 1869. See also, Pakenham, 192–3.

36 **A year later:** Dike, 189.

37 **By 1870:** Walter I. Ofonagoro, "Notes on the Ancestry of Mbanaso Okwaraozurumba Otherwise Known as Jaja of Opobo, 1821–1891," *Journal of the Historical Society of Nigeria* 9, no. 3 (December 1978), 146.

37 **"Jaja's blockade":** Dike, 193, quoting Foreign Office memo from July 11, 1870.

37 **His rivalry:** Meredith, 409. See also Falola, from p. 40.

37 **In 1882:** Pakenham, 193.

37 **London finally acquiesced:** Ikime, 64. See also, Flint, 59, and Pakenham, 199.

37 **In 1884:** Ikime, 55. See also, Pakenham, 200, and Flint, 59.

38 **Though Jaja agreed:** Dike, 215.

38 **Goldie, having by then:** Dike, 214.

38 **The General Act:** General Act of the Berlin Conference on West Africa, February 26, 1885.

38 **now officially the Oil Rivers Protectorate:** Ikime, 57.

38 **The board of directors:** Flint, 92.

38 **Company agents:** Ikime, 63.

39 **Goldie put in place:** Flint, 96.

39 **for "pacifying":** Dike, 212.

39 **ceding to the RNC:** Frederick Pedler, *The Lion and the Unicorn in Africa: The United Africa Company 1787–1931* (London: Chaucer Press, 1974), 119.

40 **On August 9, 1891:** King Jaja Death Announcement, *New York Times*, 1891. Retrieved from the *New York Times* archive.

40 **Visitors to Opobo:** Dike, 225.

40 **By 1892:** Ikime, 81–3.

40 **The end of 1894:** Pakenham, 464.

40 **On the night of:** Flint, 201, from Foreign Office memo, "Statement of Father Bubendorf" (the only European witness in Nembe).

41 **They also snuck:** Flint, 202, from Foreign Office memo. See also, Pakenham, 462 3.

41 **"Our boys fired":** Flint, 201, from Foreign Office memo.

41 **"We always looked":** Flint, 204, from Foreign Office memo.

41 **His first step:** Ikime, 62. See also, Flint, 205.

41 **"People of Brass":** Flint, 205, from Foreign Office memo.

42 **When rumors began:** Flint, 209.

42 **"Jaja was deported":** Captain MacDonald to Foreign Office, February 25, 1895, quoted in Flint, 207.

43 **Goldie, who by that point:** Flint, 3.

43 **The area claimed:** Pakenham, 650.

43 **Now they'd begun:** Flint, 243–5. See also, Pakenham, 514–22.

43 **In January of 1897:** Flint, 250–6.

43 **A year after that:** Pakenham, 514–5, 522.

44 **On January 1, 1900:** Flint, 307. According to Bank of England Inflation Calculator: 865,000 GBP in 1900 = £107,118,967.39, which equals $142,355,741.

44 **precursors to what:** Prinesha Naidoo, "Nigeria Now Tops South Africa as the Continent's Biggest Economy," *Bloomberg News*, March 3, 2020.

44 **Goldie is still:** Dike, 213.

44 **is routinely compared:** Flint, vii.

44 **having a rare:** Bo Beolens, Michael Watkins, and Michael Grayson, *The Eponym Dictionary of Reptiles* (Baltimore, MD: Johns Hopkins University Press, 2011), 103.

Chapter 2: The Flavor of Home

45 **"It is as if":** Roger Bastide, *The African Religions of Brazil: Toward a Sociology of the Interpenetration of Civilizations* (1960; reprint, Baltimore, MD: Johns Hopkins University Press, 1978), 224.

45 **It was "as if":** Mary Kingsley, *Travels in West Africa* (1897; reprint, Washington, DC: National Geographic Society, 2002), 54.

45 **we were an ocean away:** I traveled to Bahia in February of 2018.

46 **Soon after arriving:** Jean M. Hébrard, "Slavery in Brazil: Brazilian Scholars in the Key Interpretive Debates," translated by Thomas Scott-Railton, from *Translating the Americas* (Ann Arbor: The University of Michigan Center for Latin American and Caribbean Studies, 2013).

46 **When the natives:** John Thornton, *Africa and Africans in the Making of the Atlantic World, 1400–1800*, second edition (Cambridge, U.K.: Cambridge University Press, 1998), 134.

46 **By 1888:** Patrick A. Polk, Roberto Conduru, Sabrina Gledhill, and Randal Johnson, eds., *Axé Bahia: The Power of Art in an Afro-Brazilian Metropolis* (Los Angeles: Fowler Museum at UCLA, 2018), 58. See also, Slave Voyages database, https://www.slavevoyages.org. Henry Louis Gates Jr. points out in a post for The Root (https://www.theroot.com/how-many-slaves-landed-in-the-us -1790873989) that only 10.7 million of the 12.5 million that left Africa survived.

47 **Some 54 percent:** Polk et al., 58.

47 **Having made:** Case Watkins, "African Oil Palms, Colonial Socioecological Transformation and the Making of an Afro-Brazilian Landscape in Bahia, Brazil," *Environment and History*, 2015. I corresponded with Watkins over the course of writing this book. He generously provided me with his PhD thesis and with articles he has written on the topic. His first book, *Palm Oil Diaspora: Afro-Brazilian Landscapes and Economies on Bahia's Dendê Coast*, is forthcoming in June 2021 from Cambridge University Press. See also, Thornton, 156; and Judith A. Carney and Richard Nicholas Rosomoff, *In the Shadow of Slavery: Africa's Botanical Legacy in the Atlantic World* (Berkeley: University of California Press, 2009), 94.

47 **the 1642 ledger:** Adam Jones, *West Africa in the Mid-Seventeenth Century: An Anonymous Dutch Manuscript*, African Studies Association, 1994, quoted in Watkins, 22.

47 **women captives:** Thornton, 156. See also, Marcus Rediker, *The Slave Ship, A Human History* (New York: Viking, 2007), 271–2; and C. R. Boxer, "Salvador Correia de sá e Benevides and the Reconquest of Angola in 1648," *The Hispanic American Historical Review* 28, no. 4 (Nov. 1948): 492.

47 **When British forces:** H.W. Macaulay and R. Doherty, "Her Majesty's Commissioners to Viscount Palmerston, in 'Correspondence with the British Commissioners, at Sierra Leone, The Havana, and Rio de Janeiro, Related to the Slave Trade from February 2 to May 31, 1839'" (London: W. Clowes and Sons, 1839), 6.

47 **palm oil would:** Rediker, 240, 354.

48 **The region remains:** Corley and Tinker, 2.

48 **In 1699:** William Dampier, "Of the Palmberries, Physick-nuts, Mendibees, Etc, and Their Roots and Herbs, etc.," in *Dampier's Voyages*, vol. 2 (London: E. Grant Richards, 1906). Retrieved through Project Gutenberg.

48 **The word *dendê*:** Carney and Rosomoff, 198.

48 **Most of the slaves:** Thornton, 191. See also, Carney and Rosomoff, 98.

48 **In 1751:** Arquivo Histórico Ultramarino (AHU), Conselho Ultramarino, Caixa 2, Docs. 167–8, quoted in Watkins, 25.

48 **Their concern:** Corley and Tinker, 72, 75.

48 **the land was better:** Stuart B. Schwartz, "The Manumission of Slaves in Colonial Brazil: Bahia, 1684–1745," *The Hispanic American Historical Review* 54, no. 4 (Nov. 1974): 608.

48 **Beginning in 1639:** Stuart B. Schwartz, "Plantations and Peripheries, c. 1580–c. 1750," in Leslie Bethell (ed.) *Colonial Brazil* (Cambridge, U.K.: Cambridge University Press, 1987), 108. See also, Carney and Rosomoff, 126.

48 **they would often:** Carney and Rosomoff, 40. See also, Watkins, 33.

49 **Traveling through Bahia:** Johann von Spix and Karl von Martius, *Através da Bahia*, 3d ed., trans. Manoel Augusto Pirajá da Silva and Paulo Wolf (São Paulo: Companhia editora nacional, 1938), 85, quoted in Watkins, 26; and Luis Nicolau Parés, *The Formation of Candomblé: Vodun History and Ritual in Brazil* (Chapel Hill: University of North Carolina Press, 2013), 298.

49 **A Salvador-based:** Charoth doesn't speak English, and I don't speak Portuguese. I worked with a local fixer named Jon Lewis.

49 **a local *terreiro*:** Sheila S. Walker, "Everyday and Esoteric Reality in the Afro-Brazilian Candomblé," *History of Religions* 30, no. 2 (Nov. 1990). 104.

50 **the region's *quilombos*:** Carney and Rosomoff, 99. See also, Parés, 130.

50 **The complex system:** Walker, 103.

50 **devotees of *candomblé*:** Raul Lody, *Tem Dendê, Tem Axé: Etnografia do Dendezeiro* (Rio de Janeiro: Pallas, 1992), quoted in Watkins, 26.

50 **the ships trafficking:** Watkins, 38.

50 **In a letter:** Luís dos Santos Vilhena, *Bahia in the 18th Century*, 3 vols. (Salvador: Editôra Itapuã, 1969). *Compilation of News from Salvador and Brazil*, known as "The Letters of Vilhena," quoted in Watkins, 38. See also, Jeferson Bacelar, "Bahian Food in the Bitter Taste of Vilhena," *Afro-Asia*, no. 48, July/Dec 2013.

51 **officers in Bahia:** Watkins, 39–40.

51 **"Naturally, as"**: R.N. Rodrigues, *Os Africanos No Brasil* (São Paulo: Companhia Editora Nacional, 1932), quoted in Case Watkins, "Landscapes and Resistance in the African Diaspora: Five Centuries of Palm Oil on Bahia's Dendê Coast," *Journal of Rural Studies* 61 (July 2018).

51 **Among the top:** Based on my time in Salvador. See also, Elizabeth Heilman Brooke, "In Brazil's Food Capital, Tastes to Please the Gods," *New York Times*, November 4, 1992.

51 **the aforementioned *acarajé*:** Bruno Reinhardt, "Intangible Heritage, Tangible Controversies: The Baiana and the Acarajé as Boundary Objects in Contemporary Brazil," from Birgit Meyer and Mattijs van de Port, eds., *Sense and Essence: Heritage and the Cultural Production of the Real* (New York: Berghahn Books, 2018), 79.

51 **On the "street":** I visited the São Joaquim market on February 1, 2018.

53 **ending with a dish:** Carney and Rosomoff, 194.

55 **the central *terreiro* mission:** Polk et al., 29.

55 **Ayrson Heráclito:** I interviewed Heráclito via Skype on August 23, 2019.

56 **"The Atlantic":** For more on this, see Thornton, 183–205.

57 *orishas*: Parés, 41.

57 **favorite dish and color:** Reinhardt, 79.

57 **Tributes to Oshala:** Heilman Brooke, *New York Times*.

57 **The oil:** Parés, 298.

57 **Her physical manifestation:** Walker, 106. See also, Polk et al., 29.

58 **"incorporated" by spirits:** Thornton, 243. See also, Parés, 94, 106.

58 **"Any study":** Frantz Fanon, *The Wretched of the Earth* (1961; reprint, New York: Grove Press, 2004), 19.

Chapter 3: The Napoleon of Soap

60 **"This man of vast":** Edmund Morel to Vandervelde, March 29, 1911. Morel Papers, London School of Economics, F8, File 100, quoted in Brian Lewis, *"So Clean": Lord Leverhulme, Soap and Civilization* (Manchester, U.K.: Manchester University Press, 2008), 168. Translated from the French: *"Cet homme aux idées commerciales très vastes est lancé dorénavant dans l'Afrique occidentale, dans le Congo, où il peut devenir une force enorme pour le bien ou pour le mal."* I am indebted to Lewis for much of the biographical information about Lever. His wide-ranging biography looks at the man through the social and cultural context of his time.

60 **As the crow flies:** Google Maps, Douglas, Isle of Man, to Bolton: 97 miles.

60 **born in 1851:** *Port Sunlight Village*, Unilever Archives & Records Management (UARM), 1.

60 **the servants' quarters:** J.E. Flint, *Sir George Goldie and the Making of Nigeria* (Oxford, U.K.: Oxford University Press, 1960), 3.

60 **Bisected:** Brian Lewis, *The Middlemost and the Milltowns: Bourgeois Culture and Politics in Early Industrial England* (Stanford: Stanford University Press, 2001), 333–6.

60 **"Even in":** Friedrich Engels, *The Condition of the Working-Class in England in 1844* (reprint, Moscow: Panther Edition, 1969, from text provided by the Institute of Marxism-Leninism), 55.

60 **just as George Goldie:** Goldie was born in 1846, per Flint, 3.

60 **He was educated:** Frederick Pedler, *The Lion and the Unicorn in Africa: The United Africa Company 1787–1931* (London: Chaucer Press, 1974), 172.

61 **sacks of oil-palm kernels:** Martin Lynn, *Commerce and Economic Change in West Africa: The Palm Oil Trade in the Nineteenth Century* (Cambridge, U.K.: Cambridge University Press, 1997), 84–5.

62 **"People already":** UARM, LBC 8104D, Lever to G. Edward Atkinson, November 5, 1923, quoted in Lewis, 56.

62 **Business was so robust:** *Port Sunlight Village*, UARM, 1.

64 **His aim:** Birkenhead News, November 24, 1900, quoted in W.L. George, *Engines of Social Progress* (London: Adam and Charles Black, 1907), 123, and by Lewis, 99.

64 **within three years:** *The Formation of Unilever*, UARM, 4.

64 **Lever proceeded:** Lewis, 60.

65 **"At this our first":** Thomas H. Mawson, *The Life and Work of an English Landscape Architect: An Autobiography* (New York: Charles Scribner's Sons, 1927), 116, quoted in Lewis, 2.

65 **My own train:** I visited Port Sunlight in June of 2017.

66 **it still ranks:** According to the Unilever website and its Sustainable Sourcing page.

66 **on the homepage:** The homepage has since changed.

66 **letters typewritten:** From copies of various letters I read at the archives.

66 **made a baronet:** Pedler, 172.

67 **a product whose:** *Encyclopedia Britannica*, "Hippolyte Mège Mouriès."

67 **His innovation:** *The Formation of Unilever*, UARM, 4.

67 **a Dutch dairy family:** Lynn, 118.

67 **a German scientist:** American Oil Chemists Society, AOCS Library entry on Wilhelm Normann. See also, *The Formation of Unilever*, UARM, 6.

67 **a second family:** *The Formation of Unilever*, UARM, 4.

67 **In 1908:** Lewis, 11.

68 **Edwin C. Kayser:** According to the website of the United States Department of Agriculture, "Giants of the Past: The Battle Over Hydrogenation (1903–1920)." https://www.ars.usda.gov/research/publications/publication/?seqNo115=210614.

68 **"At very first glance":** Colin MacDonald, *Highland Journey* (Edinburgh and London: Moray Press, 1943), 140, quoted in Lewis, 16.

68 **Lever approached:** Pedler, 174.

68 **Having also been refused:** Jules Marchal, *Lord Leverhulme's Ghosts: Colonial Exploitation in the Congo* (London: Verso, 2008), 1.

69 **the Welsh journalist:** Pakenham, 149.

69 **Between 1879 and 1884:** Adam Hochschild, *King Leopold's Ghost: A Story of Greed, Terror, and Heroism in Colonial Africa* (New York: Houghton Mifflin, 1998), 71, 109, 110. See also, Maya Jasanoff, *The Dawn Watch: Joseph Conrad in a Global World* (New York: Penguin Press, 2017), 176, 179.

69 **At the same Berlin conference:** Jasanoff, 178. See also, Hochschild, 84–7.

69 **Leopold quickly grew rich:** Hochschild, 64.

69 **After John Boyd Dunlop's:** Hochschild, 159.

69 *Landolphia owariensis:* Royal Botanic Gardens Kew, Plants of the World Online. http://www.plantsoftheworldonline.org/taxon/urn:lsid:ipni.org:names:79672-1.

69 **The market for rubber:** Hochschild, 158–60. See also, Pakenham, 524 and 588.

69 **a first cousin:** Pakenham, 12.

70 **Leopold had declared:** Hochschild, Pakenham, and Jasanoff.

71 **who'd since left:** Hochschild, 206.

71 **he had recently made:** Marchal, 2.

71 **In 1911:** Marchal, 1; Lewis, 171.

71 **The thirty-three-year lease:** Pedler, 175; Lewis, 171. Converted from 25 centimes per hectare.

71 **the most promising:** Marchal, 2.

72 **"Lever is a man":** London School of Economics, Morel Papers, F8, File 100, Morel to Vandervelde, March 29, 1911.

73 **he'd shipped one thousand:** Marchal, 2.

73 **a single bar:** Hochschild, 266. See also, Marchal, 2.

73 **Lever voyaged:** Pedler, 180. See also, Marchal, 2.

73 **"Going up that river":** Joseph Conrad, *Youth/Heart of Darkness/The End of the Tether* (London: Penguin Books, 1995), 88.

73 **writer was sick:** Jasanoff, 180–200.

73 **"saw a camp[in]g place":** Hochschild, 144; Jasanoff, 191.

73 **"an immense snake":** Conrad, 54.

73 **"all healthy and strong":** Diary entry, January 1, 1913, quoted in Lewis, 173.

74 **"The native material":** W.P. Jolly, *Lord Leverhulme* (London: Constable, 1976), 125, quoted in Lewis, 175.

74 **"It is a well-known fact":** UARM, LBC 8104A, Lever to H.R. Greenhalgh, March 21, 1916.

74 **local agents took to raiding:** Marchal, 6, 14–18.

75 **"Obviously our only right":** UARM, LBC 8104A, Lever to H.R. Greehalgh, March 8, 1918, quoted in Lewis, 170.

75 **"The trade that has taken":** UARM, LBC 118A, Lever to A.P. van Geelkerken, August 23, 1915, quoted in Lewis, 199.

75 **he would profit:** *The Formation of Unilever*, UARM, 6. See also, Pedler, 181.

75 **Between 1914 and 1918:** Susan M. Martin, *Palm Oil and Protest: An Economic History of the Ngwa Region, South-Eastern Nigeria, 1800–1980* (Cambridge, U.K.: Cambridge University Press, 1988), 57. See also, Lewis, 10.

75 **palm oil shot:** Martin, 57.

76 **"The Company is finding":** Marchal, 17–18.

76 **The HCB:** Marchal, 27–9.

76 **He went on:** Marchal, 29.

77 **Lejeune described:** Marchal, 30–5.

77 **Fewer than half:** Unilever's World, Anti Report No. 11, Counter Information Services, London 1974, p. 2. See also, John Tanner, "Simply . . . Unilever's History," *New Internationalist*, issue 172, June 1987. https://newint.org /features/1987/06/05/simply.

77 **"No man":** Charles Wilson, *The History of Unilever: A Study in Economic*

Growth and Social Change, vol. I (London: Cassell and Co., 1954), 150, quoted in Lewis, 121.

77 **"Bosses in Africa":** Marchal, 39–40.

78 **"At present":** Marchal, 62.

78 **he had purchased:** *The Formation of Unilever*, UARM, 12. See also, Lewis, 188.

78 **paying more than £8 million:** Lewis, 200.

79 **Goldie summoned:** Flint, 318.

79 **Lever took a lease:** *Unilever House*, UARM, 4–5.

79 **the *Daily Standard* described:** *Unilever House*, UARM, 5, 7.

79 **second and final visit:** *United Africa Company*, UARM, 7. See also, Pedler, 187.

79 **in a *"Cabine de Luxe"*:** UARM, LBC 4506, Lever to Annie and D'Arcy Lever, November 12, 1924, quoted in Lewis, 177.

79 **Its African workforce:** Lever, Leverhulme, 310–11; Fieldhouse, Unilever Overseas, 507, quoted in Lewis, 178.

79 **"On the foreshore":** UARM, LBC 4271, Lever to Myrtle Huband, October 27, 1924, quoted in Lewis, 177.

80 **"Our grievances":** Martin, *Palm Oil and Protest*, 106, 116. See also, Marchal, 219.

80 **Among the worst:** Charles Sikitele Gize, "The Revolt of the Pende," *Cercle d'Art des Travailleurs de Plantation Congolaise*, Els Roelandt and Eva Barois de Caevel, eds. (Berlin: Sternberg Press, 2017), 247–75. See also, Marchal, 148–51.

80 **"It can be said":** Marchal, 166.

81 **In 1929:** *United Africa Company*, UARM, 3.

81 **what *The Economist*:** *The Formation of Unilever*, UARM, 3.

81 **In 1939:** Marchal, 211–8.

82 **His patronage:** Harley Williams, *Men of Stress: Three Dynamic Interpretations, Woodrow Wilson, Andrew Carnegie, William Hesketh Lever* (London: Jonathan Cape, 1948).

82 **"In later years":** William Hulme Lever [second Viscount Leverhulme], Viscount Leverhulme (Boston, MA, and New York: Houghton Mifflin, 1927), 67, quoted in Lewis, 2.

Chapter 4: Playboys of the South China Sea

83 **"This jungle"**: Henri Fauconnier, *The Soul of Malaya* (1931; reprint, Singapore: Archipelago Press, 2003), 21.

83 **Standing on the deck:** Biographical information here mostly from Roland Fauconnier, who provided the biographical notes for the 2003 edition of *The Soul of Malaya.*

84 **"a thick, warm, yellow odour"**: Henri Fauconnier, 114.

84 **Three years earlier:** According to the website of the Raffles Hotel. http://www.rafflessingapore.com/raffles-history.

84 **Three decades earlier:** Charles C. Mann, "Why We (Still) Can't Live Without Rubber," *National Geographic*, December 2015.

84 **With European financiers:** Henry S. Barlow, *The Malaysian Plantation Industry: A Brief History to the Mid-1980s* (Kuala Lumpur: Arabis, 2018). Accessed online.

84 **Whereas in 1890:** T.R. McHale, *Rubber and the Malaysian Economy* (Sendirian Berhad, MPH Publications, 1967).

84 **it was also:** Alec Gordon, "Contract Labour in Rubber Plantations: Impact of Smallholders in Colonial Southeast Asia," *Economic and Political Weekly*, March 10, 2001.

85 **In 1905:** Encyclopedia.com, "Harrisons & Crosfield."

85 **the explosion:** For general history, see "The International Natural Rubber Market, 1870–1930," Economic History Association.

85 **the southern Indians:** *Encyclopedia Britannica*. See also, "A History of Indentured Labor Gives 'Coolie' Its Sting," *NPR Code Switch*, November 25, 2013.

85 **"So much to think about"**: R. Fauconnier, notes from *The Soul of Malaya*, 182. I converted 600 hectares to 1,482 acres.

86 **"with gigantic trees"**: Madelon Lulofs, *White Money* (London and New York: The Century Co., 1933), 8.

86 **In 1909:** Socfindo Sustainability Report 2018, p. 7. https://www.socfin.com/sites/default/files/2019-09/Socfindo%20GRI%20Report_2018_1.pdf https://www.socfin.com/en/key-dates.

86 **They marked:** Valeria Giacomin, "The Transformation of the Global Palm Oil Cluster: Dynamics of Cluster Competition Between Africa and Southeast Asia (c. 1900–1970)," *Journal of Global History* 13, no. 3 (November 2018): 384. See also, Susan M. Martin, *The UP Saga* (Copenhagen: Nordic Institute of Asian Studies Press, 2003), 49. I am indebted to Martin for much

of the early history of the industry in Malaysia and, specifically, for the details about United Plantations.

86 **The oil palm:** R.H.V. Corley and P.B. Tinker, *The Oil Palm*, 5th ed. (West Sussex, U.K.: John Wiley & Sons, 2016), 6. See also, "Celebrating 100 Years of Malaysian Palm Oil," *New Straits Times*, May 19, 2017.

88 **by 1915:** John H. Drabble, "The Economic History of Malaysia," *EH.Net Encyclopedia*, Robert Whaples, ed., July 31, 2004. http://eh.net/encyclopedia /economic-history-of-malaysia.

88 **The couple:** Bernard Fauconnier, *La Fascinante Existence d'Henri Fauconnier* (Saint-Malo: Gérard Desquesses, 2003), 116–117.

88 **By the early 1920s:** Martin, *The UP Saga*, 51.

88 **Oil-palm plantations:** Martin, *The UP Saga*, 52. I converted 31,000 hectares to 76,602 acres; 3,400 hectares to 8,401 acres; and 20,500 hectares to 50,656 acres. See also, Corley and Tinker, 6: Sumatra, 31,600 hectares. Malaya: 3,350.

88 **More and more:** Giacomin, 386.

89 **In the meantime:** Martin, *The UP Saga*, 68.

89 **Shipments bound:** Giacomin, 389.

89 **Among the other:** All from Martin, *The UP Saga*.

90 **Though the new plot:** Martin, *The UP Saga*, 25–46. Converted from 27 kilometers, per Martin (16.777 miles).

90 **It was a random run-in:** Martin, *The UP Saga*, 60–3.

91 **Graham Greene:** Jerry Hopkins, *Romancing the East: A Literary Odyssey from the Heart of Darkness to the River Kwai* (Clarendon, VT: Tuttle Publishing, 2013), 65.

91 **also the Hungarians:** Ladislao Székely, *Tropic Fever: The Adventures of a Planter in Sumatra* (New York: Harper & Brothers, 1937); Madelon Lulofs, *White Money: A Novel of the East Indies* (London: The Century Company, 1933), *Rubber: The 1930s Novel Which Shocked European Society* (Oxford, U.K.: Oxford University Press, 1933), and *Coolie* (Oxford, U.K.: Oxford University Press, 1936).

92 **"This frenzy":** Fauconnier, 161.

92 **Having fled war:** Sunil S. Amrith, *Crossing the Bay of Bengal: The Furies of Nature and the Fortunes of Migrants* (Boston: Harvard University Press, 2013). See also, Martin, *The UP Saga*, 58.

92 **"From a labour":** Charles Hirschman, "The Making of Race in Colonial Malaya," *Sociological Forum* 1, no. 2 (Spring 1986): 330–61.

92 **"I hoe all day":** Marina Carter and Khal Torabully, *Coolitude: An Anthology of the Indian Labour Diaspora* (London: Anthem Press, 2002), 22.

93 **By 1934:** Martin, *The UP Saga*, 67.

93 **when officials:** The British National Archives, London, "Palm Oil Expedition to Sumatra, 1926," quoted in Giacomin.

93 **In short:** Corley and Tinker, 19.

93 **in Sumatra:** Corley and Tinker, 6. 1938, Sumatra: 92,000 hectares = 227,000 acres. Malaysia: 20,000 = 49,000 acres.

93 **in Malaya:** "East Sumatra's Growth: Planter and Peasant, Population and Communications," Chapter IV from *Planter and Peasant: Colonial Policy and the Agrarian Struggle in East Sumatra 1863–1947* (Leiden: Brill Publishing, 1978).

94 **several hundred workers:** Barlow.

94 **Returning to their:** Martin, *The UP Saga*, 90–1.

94 **Among the places:** I traveled to Sandakan in October of 2016.

94 **The airstrip:** Max Hastings, "The Untold Story of the Sandakan Death Marches, by Paul Ham," *The Sunday Times*, July 21, 2013.

95 **A continent away:** According to the Nutella website. See also, Noah Kirsch, "The Nutella Billionaires: Inside the Ferrero Family's Secret Empire," *Forbes*, June 26, 2018.

95 **Prior to the outbreak:** Martin, *The UP Saga*, 49–50. See also, Barlow, section d.

96 **Such urbanites:** Christopher Hale, *Massacre in Malaya, Exposing Britain's My Lai* (Gloucestershire: The History Press, 2013), 381. See also, Ronnie Tan, "Civilians in the Crossfire: The Malayan Emergency," *BiblioAsia* (National Library Singapore) 15, no. 3 (2019).

97 **a golf club:** The club, Fraser's Hill, was "near Raub in Pahang." According to Google Maps, Fraser's Hill is 97 kilometers from KL city center. 97 km = 60 miles.

97 **An earlier massacre:** Hale, 391–404. See also, Mark Townsend, "Revealed: How Britain Tried to Legitimise Batang Kali Massacre," *The Guardian*, May 5, 2012.

97 **In 1955:** Geoffrey Jones and Judith Wale, "Diversification Strategies of British Trading Companies: Harrisons & Crosfield," *Business History* 41, no. 2 (1999).

97 **Hallet's Socfin:** Martin, *The UP Saga*, 149.

97 **Back in the Congo:** Martin, *The UP Saga*, 184–8.

97 **"Colors, moisture":** Joan Didion, *Democracy* (1984; reprint, New York: Random House, 1995), 16.

98 **"We were sitting":** Didion, 228.

98 **descending over Kuala Lumpur:** I visited Kuala Lumpur in November of 2014, October of 2016, and March of 2017.

98 **seven-million-plus citizens:** According to World Population Review, 7,458,015 people. https://worldpopulationreview.com/countries/malaysia-population.

98 **it established something:** Derek Byerlee, Walter P. Falcon, and Rosamond L. Naylor, *The Tropical Oil Crop Revolution: Food, Feed, Fuel, & Forests* (Oxford, U.K.: Oxford University Press, 2017), 25.

99 **FELDA moved more deeply:** Barlow, section d.

100 **By 1966:** Martin, *The UP Saga*, 199–204.

100 **In the late 1950s:** Nita G. Forouhi, Ronald M. Krauss, Gary Taubes, and Walter Willett, "Dietary Fat and Cardiometabolic Health: Evidence, Controversies, and Consensus for Guidance," *BMJ*, June 13, 2018, Box 1.

100 **By 1969:** Ruth DeFries, *The Big Ratchet: How Humanity Thrives in the Face of Natural Crisis* (New York: Basic Books, 2014), 193.

101 **grew more than twenty-five-fold:** I have relied on stats from Index-Mundi, https://www.indexmundi.com/agriculture/?country=my&commodity=palm-oil&graph=production. 1964: 151,000 metric tons. 1984: 3,817,000 metric tons.

101 **But a Scot:** Corley and Tinker, 9, 46–7. See also, "In Memorium: Datuk Leslie Davidson (1931–2019)," *Global Oils and Fats* no. 2 (June 2019).

102 **The larger ones:** Byerlee, Falcon, and Naylor, 30–5.

102 **processes some thirty thousand tons:** According to a 2017 Audit Report submitted to the Roundtable on Sustainable Palm Oil. https://rspo.org/uploads/default/pnc/PPB_Terusan_POM_draft_RSPO_Audit_Report_2017_final27102017_PS_revised.pdf.

102 **more than 450 mills:** "Malaysian Palm Oil Industry Aims for 70 Percent Certification by February 2020," *Oils & Fats International*, November 19, 2019.

102 **cover some 14 million acres:** According to the Center for International Forestry Research, or CIFOR. https://www.cifor.org. See also, A. Ananthalakshmi, "Palm Oil to Blame for 39 Percent of Forest Loss in Borneo Since 2000—Study," *Reuters*, September 19, 2019.

103 **sterilized the fruits:** Corley and Tinker, 468–9.

104 **the media representative:** Iris Chan Suet Yeng, corporate communications.

105 **The resulting "RBD":** Corley and Tinker, 28, 479.

105 **further "fractionate"**: Technicians at the refinery. See also, Martin, 237.

105 **some two hundred names**: Margie Mason and Robin McDowell, "Palm Oil Labor Abuses Linked to Top Brands, Banks," *Associated Press*, September 24, 2020.

105 **20.5 million metric tons**: According to statistics from the USDA. https://ipad.fas.usda.gov/highlights/2019/05/malaysia/index.pdf.

105 **worth some $9 billion**: The Observatory of Economic Complexity, https://oec.world/en/profile/hs92/31511.

105 **20 million acres**: Global Forest Watch. I converted 8.12 million hectares to 20.06 million acres.

106 **On his deathbed**: Mohamad Rashidi Pakri and Arndt Graf, eds., *Fiction and Faction in the Malay World* (Newcastle Upon Tyne: Cambridge Scholars Publishing, 2012), 29.

Chapter 5: Silent Summers

109 **"There was a strange stillness"**: Rachel Carson, *Silent Spring* (1962; reprint, New York: Houghton Mifflin Harcourt Publishing, 2002), 2.

110 **By the time I arrived**: I visited Sumatra in May of 2015, February and November of 2016, and July of 2018.

110 **more than 75 percent**: BirdLife International, Data Zone. http://datazone.birdlife.org/sowb/casestudy/in-current-global-markets-oil-palm-plantations-are-valued-more-highly-than-ancient-forest.

110 **Long prized**: N.J. Collar, "Helmeted Hornbills: Rhinoplax vigil and the Ivory Trade: The Crisis That Came Out of Nowhere," *BirdingASIA* 24 (2015): 12–17. See also, Apriadi Gunawan, "Poachers Shift to Hornbills as Rare Animals Decline," *Jakarta Post*, June 17, 2015; and Bambang Muryanto, "Poaching of Endangered Hornbill Continues Amid Global Demand," *Jakarta Post*, February 26, 2016.

110 **The wildlife-trade watchdog**: "Mapping seizures to aid conservation of imperiled Helmeted Hornbill," website of TRAFFIC, based in Cambridge, U.K. Posted May 16, 2017.

110 **In 2016**: "Traders of One of Indonesia's Most Hunted Bird Species Arrested," *WCS Newsroom*, June 18, 2015.

111 **Yokyok "Yoki" Hadiprakarsa**: I have communicated with Hadiprakarsa by email since April of 2016.

112 **Though the Dutch**: Joshua Oppenheimer, "Show of Force: Film, Ghosts, and Genres of Historical Performance in the Indonesian Genocide," PhD thesis, 2004, University of the Arts London, 13.

112 **The initiatives:** Ann Laura Stoler, *Capitalism and Confrontation in Sumatra's Plantation Belt, 1870–1979* (Ann Arbor: University of Michigan Press, 1995), 2–3.

112 **the PKI had become:** Oppenheimer, 15.

112 **In October of 1965:** Geoffrey B. Robinson, *The Killing Season: A History of the Indonesian Massacres, 1965–66* (Princeton, NJ: Princeton University Press, 2018). I am reliant on Robinson's book for much of the history in this chapter.

112 **which the CIA:** "Indonesia—The Coup That Backfired," Research Study, December 1968, The CAESAR, POLO, and ESAU Papers, United States Central Intelligence Agency.

113 **Documentary evidence:** Margaret Scott, "The Indonesian Massacre: What Did the U.S. Know?" *New York Review of Books*, November 2, 2015.

113 **"Every Saturday night":** Robinson, 126.

113 **60 percent of its citizens:** Walter P. Falcon, "Food Security for the Poorest Billion: Policy Lessons from Indonesia," in Rosamond L. Naylor (ed.), *The Evolving Sphere of Food Security* (Oxford, U.K.: Oxford University Press, 2014), 33.

113 **Whereas in 1965:** Joanne C. Gaskell, "The Role of Markets, Technology and Policy in Generating Palm Oil Demand in Indonesia," *Bulletin of Indonesian Economic Studies* 51, no. 1 (2015-03-30): 1, 12.

114 **The folks on the receiving end:** "New Order Forestry Policy and the Roots of the Crisis," Human Rights Watch, 2003. https://www.hrw.org/reports /2003/indon0103/Indon0103-02.htm.

114 **have lost tens of thousands of acres:** "When We Lost the Forest, We Lost Everything," Human Rights Watch, 2019. https://www.hrw.org/report /2019/09/23/when-we-lost-forest-we-lost-everything/oil-palm-plantations -and-rights-violations. See also, Barbara Beckert, Christoph Dittrich, and Soeryo Adiwibowo, "Contested Land: An Analysis of Multi-Layered Conflicts in Jambi Province, Sumatra, Indonesia," *Austrian Journal of South-East Asian Studies* 7, no. 1 (2014): 75–92; and Hannah Beech, "Their Land Defiled, Forest People Swap Flower Worship for Quran and Concrete," *New York Times*, October 14, 2018.

115 **In 2018:** "When We Lost the Forest, We Lost Everything," Human Rights Watch, 2019.

115 **The 2019 U.N. report:** *2019 Global Assessment Report on Biodiversity and Ecosystem Services*, coordinated by the Bonn-based Intergovernmental Science-Policy Platform on Biodiversity and Ecosystem Services (IPBES). See also, Rachel Nuwer, "Mass Extinctions Are Accelerating," *New York Times*, June 1, 2020.

115 **tropical rainforests, which:** E.O. Wilson and Frances M. Peter, eds, *Biodiversity.* Chapter 3: "Tropical Forests and their Species Going, Going . . . ?" by Norman Myers (Washington, DC: National Academies Press, 1988). See also, Rhett A. Butler, "Ten Rainforest Facts for 2020," *Mongabay,* July 12, 2020.

115 **the Leuser Ecosystem:** Haka Indonesia. https://www.haka.or.id/?page_id=1008.

115 **Home to 382 bird:** Rainforest Action Network, "Leuser Watch," September 2019. See also Paul Hilton, "Leuser Ecosystem: One of Most Biodiverse Places on Earth, in Pictures," *The Guardian,* September 28, 2017.

116 **In addition:** John MacKinnon and Karen Phillipps, *A Field Guide to the Birds of Borneo, Sumatra, Java, and Bali: The Greater Sunda Islands* (Oxford, U.K.: Oxford University Press, 1993).

116 **only 4.5 million acres:** Rainforest Action Network. https://www.ran .org/issue/leuser. See also, Junaidi Hanafiah, "Restoring Sumatra's Leuser Ecosystem, One Small Farm at a Time," *Mongabay,* September 30, 2019.

116 **At $6,000 a kilogram:** "The Bird That's More Valuable Than Ivory," *BBC,* October 12, 2015.

117 **Hong Kong shops:** "Seeing Red: The Often Hidden Colour of Wildlife Contraband," Environmental Investigation Agency, January 26, 2015.

117 **the incursions:** Darren Naish, "The Ecology and Conservation of Asian Hornbills: Farmers of the Forest," *Historical Biology,* September 2014.

117 **I'd landed:** I traveled throughout the Leuser with Putra from February 27 to March 2, 2016.

118 **a young father:** "Indonesian Man's Body Found Inside Python—Police," *BBC,* March 29, 2017.

120 **"We didn't do":** I interviewed Ngatimen on March 1, 2016.

122 **fewer than three thousand:** BirdLife International. http://datazone .birdlife.org/species/factsheet/rhinoceros-hornbill-buceros-rhinoceros.

123 **In 2008:** All information and stats according to the IUCN. https://www .iucnredlist.org.

123 **Ian Singleton:** I first met Singleton in February of 2016 and spent time with him at his office, at the SOCP quarantine, and at Roland's in November of 2016. I learned while fact-checking this book that he had recently divorced.

123 **a funky German joint:** After a long run, Roland's recently closed its doors.

126 **many of the richest:** Forbes staff, "Almost Half of Tycoons on Forbes Indonesia Rich List See Fortunes Rise," *Forbes,* December 4, 2019. Widjaja (#2), Salim (#6), Karim (#11), Sitorus (#13), Sampoerna (#14).

127 **A hotelier friend:** Given his contacts in the industry, he asked that I not identify him by name.

128 **Jennifer Draiss:** I spent time with Draiss at the quarantine in November of 2016.

134 **A 2019 study:** Jie-Sheng Tan-Soo and Subhrendu K. Pattanayak, "Seeking natural capital projects: Forest fires, haze, and early-life exposure in Indonesia," *Proceedings of the National Academy of Sciences* 116, no. 12: 5239–45.

Chapter 6: Caravan Dreams

136 **"It was there":** Gabriel García Márquez, *One Hundred Years of Solitude* (1967; reprint, New York: Harper Perennial, 2006), 301–2.

136 **Walter Banegas:** I interviewed Banegas at his home on December 3, 2018.

137 **run by a Honduran company:** RSPO ACOP report submission for 2018, https://www.rspo.org/members/1010/grupo-jaremar.

139 **workers in Southeast Asia:** Staff writer, "Plantation Worker Dies from Electrocution in Freak Accident," *The Star*, June 20, 2013.

139 **two laborers:** Lubulwa Henry, "Two Oil Palm Workers in Kalangala Electrocuted," *Uganda Radio Network*, September 20, 2019.

139 **oil-palm plantations:** According to the World Bank. I converted 190,000 hectares to 469,500 acres.

139 **In the early years:** For the history of agriculture in Honduras, I have relied on Dan Koeppel, *Banana: The Fate of the Fruit that Changed the World* (New York: Penguin, 2009); Tanya M. Kerssen, *Grabbing Power: The New Struggles for Land, Food and Democracy in Northern Honduras* (Oakland, CA: Food First Books, 2013); and Dana Frank, *The Long Honduran Night: Resistance, Terror, and the United States in the Aftermath of the Coup* (Chicago, IL: Haymarket Books, 2018).

140 **By the 1930s:** Grace Livingstone, *America's Backyard: The United States and Latin America from the Monroe Doctrine to the War On Terror* (London: Zed Books, 2009), 32.

140 **"Through bribery":** Stephen C. Schlesinger and Stephen Kinzer, *Bitter Fruit: The Untold Story of the American Coup in Guatemala* (New York: Doubleday, 1982). See also, Nicholas Stein, "Yes, We Have No Profits: The Rise and Fall of Chiquita Banana," *Fortune*, November 26, 2001.

140 **The Chiquita name:** Koeppel, 116.

140 **In 1952:** Koeppel, 126–7, 130–1, 170, 255.

141 **built research centers:** Corley and Tinker, 22. See also, Koeppel, 104.

141 **Such "American oil palm":** Michael Taussig, *Palma Africana* (Chicago: The University of Chicago Press, 2018).

141 **some 37,000 acres:** According to its most recent filing with the RSPO. https://www.rspo.org/members/1010/grupo-jaremar. I converted 14,906 hectares to 36,833 acres.

141 **Hershey's:** According to Hershey's Traceability Report. https://www.thehersheycompany.com/content/dam/corporate-us/documents/pdf/hershey-h1-2018-traceability.pdf.

141 **PepsiCo:** According to PepsiCo's mill list. https://www.pepsico.com/docs/album/esg-topics-policies/pepsico-mill-list-2019.pdf?sfvrsn=a40f742b_4.

141 **Grupo Bimbo:** According to Grupo Bimbo's mill list: https://grupobimbo.com/sites/default/files/Grupo_Bimbo_2018_Clean_Mill_List_Jan_2020.pdf.

141 **a labor-rights activist:** I traveled with Gabby Rosazza, then with the DC-based International Labor Rights Forum, in December of 2018.

142 **the hotel:** I stayed at the Gran Hotel Sula.

142 **it's a situation:** Alexander Main, "The U.S. Militarization of Central America and Mexico," NACLA, June 17, 2014.

142 **oil-palm plantations:** Steven E. Sesnie, Beth Tellman, David Wrathall, Kendra McSweeney, Erik Nielsen, Karina Benessaiah, Ophelia Wang, and Luis Rey, "A Spatio-Temporal Analysis of Forest Loss Related to Cocaine Trafficking in Central America," *Environmental Research Letters*, May 2017.

143 **Petén region:** I reported from the Petén in August of 2017.

143 **In the early 1970s:** "Lower Aguan in Honduras and the Deadly Battle Over Land Rights," Carnegie Council for Ethics in International Affairs, June 2014.

143 **a few wealthy businessmen:** Kerssen, 92.

143 **some 29 percent:** "A Scorecard of the Latin American Palm Oil Sector," Forest Heroes, 2018. https://forestheroes.com/wp-content/uploads/2018/06/Behind-the-Global-Curve-A-Scorecard-of-the-Latin-American-Palm-Oil-Sector.pdf.

143 **fifty-three murders:** Sasha Chavkin, "Bathed in Blood: World Bank's Business Arm Backed Palm Oil Producer Amid Deadly Land War," *International Consortium of Investigative Journalists*, June 9, 2015. See also, "Death Valley: The Land War Gripping Honduras," *The Irish Times*, May 8, 2015.

144 **I'll call Carlos:** He asked that I not use his name due to the security situation in the region.

145 **Among them is an insecticide:** Sonia Mejia, corporate manager for social responsibility and communications at Grupo Jaremar, told me in an

email that the company stopped using Lorsban in March of 2020. However, Ahraxa-tzel Mayorga, lead organizer for the workers' union, said that the workers have neither been told of a change nor offered instruction on applying new chemicals.

145 **chlorpyrifos has been linked:** According to the Pesticide Action Network International, 2019 List of Highly Hazardous Pesticides. http://files .panap.net/resources/Consolidated-List-of-Bans-Explanatory.pdf.

145 **In 2015:** Patricia Cohen, "Roundup Maker to Pay $10 Billion to Settle Cancer Suits," *New York Times,* June 24, 2020. In 2020, the United States Environmental Protection Agency concluded that glyphosate was not a carcinogen.

146 **there are no bathing facilities:** Jaremar spokesperson Sonia Mejia told me in an email that "there are showers for staff," though she declined to say how many or offer any proof. Union organizer Mayorga disputes the claim that plantation workers have access to showers.

147 **The latter film:** Joshua Oppenheimer, "Show of Force: Film, Ghosts, and Genres of Historical Performance in the Indonesian Genocide," PhD thesis, 2004, University of the Arts London, 45–7, 82.

147 **forty-six countries:** Pesticide Action Network, "PAN International Consolidated List of Banned Pesticides," pan-international.org. See also, "Stop Paraquat in Palm Plantations," blog, PAN, September 11, 2020.

147 **generally women:** "The Great Palm Oil Scandal," Amnesty International, 2016, 9.

147 **a 2016 report:** "The Great Palm Oil Scandal," Amnesty International, 2016, 9.

147 **among whose major shareholders:** According to the company's website. https://www.wilmar-international.com/about-us/corporate-profile.

147 **"will be phased out":** Jaremar spokesperson Sonia Mejia told me that paraquat had been phased out in 2009 but declined to explain the contradictory OPIC statement from five years later.

148 **A woman named Yohanna:** "The Great Palm Oil Scandal," Amnesty International, 74–8.

149 **"They told us":** Wudan Yan, "I've Never Been Normal Again: Indonesian Women Risk Health to Supply Palm Oil to the West," *STAT News,* April 2017.

149 **Female plantation workers:** "Women, Tree Plantations, and Violence: Building Resistances," World Rainforest Movement, Bulletin 236, March 2018.

149 **In a November 2020:** Margie Mason and Robin McDowell, "Rape, Abuses in Palm Oil Fields Linked to Top Beauty Brands," *Associated Press,* November 18, 2020.

149 **A 2015 article:** Syed Zain Al-Mahmood, "Palm-Oil Migrant Workers Tell of Abuses on Malaysian Plantations," *Wall Street Journal*, July 26, 2015.

151 **filed a complaint:** "Petition to exclude palm oil and palm oil products manufactured 'wholly or in part' by forced labor in Malaysia by FGV Holdings Berhad," August 15, 2019 letter to John P. Sanders, Acting Commissioner, U.S. Customs and Border Protection, U.S. Department of Homeland Security, from ILRF, RAN, and Sum of Us.

151 **Tariff Act of 1930:** "Activists Urge U.S. Customs to Ban Palm Oil Imports from Malaysia's FGV," *Reuters*, August 15, 2019.

151 **an investigation by the Associated Press:** Margie Mason and Robin McDowell, "US Says It Will Block Palm Oil from Large Malaysian Producer," *AP*, September 30, 2020.

151 **plantations that supply Wilmar:** "The Great Palm Oil Scandal," Amnesty International, 6, 24, 38.

152 **three plantations:** "A Dirty Investment: European Development Banks' Link to Abuses in the Democratic Republic of Congo's Palm Oil Industry," Human Rights Watch, November 2019, 1.

153 **"At first I thought":** "A Dirty Investment," annex, Part I, footnotes 11, 15, and 10.

153 **A systematic review:** Nuruly Myzabella, Lin Fritschi, Nick Merdith, Sonia El-Zaemey, HuiJun Chih, and Alison Reid, "Occupational Health and Safety in the Palm Oil Industry: A Systematic Review," *International Journal of Occupational and Environmental Medicine*, July 2019.

153 **the Congolese laborers described:** "A Dirty Investment," Part I, "Skin Problems"; Part I, footnote 25; Part I, "Inadequate Medical Care and Monitoring."

154 **Their company-issued overalls:** "A Dirty Investment," footnote 189; Part III.

155 **One mother of six:** "A Dirty Investment," Part III, "Abusive Employment Practices and Extreme Poverty Wages"; footnote 148; Part III, footnote 191.

155 **In his 1967 novel:** Gabriel García Márquez, *One Hundred Years of Solitude* (1967; reprint, New York: Harper Perennial, 2006).

155 **"I have the honor":** Koeppel, 88–9.

156 **In October of 2017:** "Honduras Report: Freedom of Association and Democracy," Solidarity Center, 2017–2018.

156 **Yarleni Ortéz Mejía:** I met with Mejía in her home in December of 2018.

156 **from January 2015:** Solidarity Center report.

157 **prompted fifty-five congressional representatives:** "Schakowsky, Grijalva, Levin Lead Letter Calling on Honduras to Halt Labor Violations," press release from U.S. Representatives Jan Schakowsky (D-IL), September 6, 2019.

157 **found themselves worse off:** Max Radwin, "It's Getting Worse: National Parks in Honduras Hit Hard by Palm Oil," *Mongabay*, April 2019.

158 **increasing numbers:** Georgina Gustin, "Ravaged by Drought, a Honduran Village Faces a Choice: Pray for Rain or Migrate," *Inside Climate News*, July 8, 2019.

158 **her father said:** Jeff Abbott and Sandra Cuffe, "Palm oil industry expansion spurs Guatemala indigenous migration," Al Jazeera, February 6, 2019.

158 **Maria Margarita Ivanez:** I interviewed Ivanez in the Petén region near her home in August of 2017.

158 **Marcelino Flores:** I interviewed Flores at his home in December of 2018.

Chapter 7: The World Is Fat

160 **"People are fed":** Wendell Berry, *Sex, Economy, Freedom & Community: Eight Essays* (New York: Pantheon, 2014), 7.

160 **Dr. Anoop Misra:** I spent time with Misra in his New Delhi office on October 24, 2017.

161 **Shubhra Atrey:** I interviewed Atrey in her office on October 24, 2017.

161 **A 2017 study:** "Health Effects of Overweight and Obesity in 195 Countries over 25 Years," *New England Journal of Medicine* 377, no. 1 (July 6, 2017).

161 **India has more:** Bee Wilson, *The Way We Eat Now: How the Food Revolution Has Transformed Our Lives, Our Bodies, and Our World* (New York: Basic Books, 2019), 60.

161 **"We have more":** Matt Richtel, "More Than 10 Percent of World's Population Is Obese, Study Finds," *New York Times*, June 12, 2017.

162 **global production:** Figures according to the FAO. http://www.fao.org/faostat/en/#compare.

162 **The growth:** Derek Byerlee, Walter P. Falcon, and Rosamond L. Naylor, *The Tropical Oil Crop Revolution: Food, Feed, Fuel, & Forests* (Oxford, U.K.: Oxford University Press, 2017), 2, 19.

162 **"It is expected":** Edy Sarif, "Huge Opportunities in Palm Oil," *The Star*, March 7, 2012.

162 **have added far more:** Corinna Hawkes, "Uneven Dietary Development: Linking the Policies and Processes of Globalization with the Nutrition

Transition, Obesity, and Diet-Related Chronic Diseases," *Global Health*, March 2006.

162 **Between 1961 and 2009:** Figures according to the International Center for Tropical Agriculture, or CIAT, "The Changing Global Diet." https://ciat .cgiar.org/the-changing-global-diet/crop-trends.

162 **by 278 calories:** Byerlee, Falcon, and Naylor, 106.

163 **Between 1980 and 2000:** Hawkes.

163 **Just as farm policies:** Michael Pollan, *The Omnivore's Dilemma: A Natural History of Four Meals* (New York: Penguin, 2006).

163 **Over the past decade:** Bee Wilson, "How Ultra-Processed Food Took Over Your Shopping Cart," *The Guardian*, February 12, 2020.

163 **In 2018:** Bernard Srour, "Consumption of Ultra-processed Foods and Cancer Risk: Results from NutriNet-Santé Prospective Cohort," *BMJ*, January 10, 2018.

164 **studies have shown:** Geng Zong, Yanping Li, Anne J. Wanders, Marjan Alssema, Peter L. Zock, Walter C. Willett, Frank B. Hu, and Qi Sun, "Intake of Individual Saturated Fatty Acids and Risk of Coronary Heart Disease in US Men and Women: Two Prospective Longitudinal Cohort Studies," *BMJ*, October 2016. See also, Ye Sun, Nithya Neelakantan, Yi Wu, Rashmi Lote-Oke, An Pan, and Rob M. van Dam, "Palm Oil Consumption Increases LDL Cholesterol Compared with Vegetable Oils Low in Saturated Fat in a Meta-Analysis of Clinical Trials," *Journal of Nutrition* 145, no. 7 (July 2015).

164 **"Trans fats carry":** I spoke with Popkin by phone on May 16, 2018, and we emailed over several weeks.

165 **at $694 a metric ton:** Figures according to the World Bank commodity sheet. https://knoema.com/WBCPD2015Oct/world-bank-commodity-price -data-pink-sheet-monthly-update?commodity=1000960&measure=1000000. They obviously fluctuate with time.

165 **"Even in developing countries":** I spoke with Qi Sun by phone on July 6, 2017.

165 **Accompanied by a researcher:** I was with Suparna Ghosh-Jerath, PhD.

166 **priced at 90:** Though prices fluctuate, palm oil is typically substantially cheaper than the others.

166 **less than $2,000:** According to the World Bank.

167 **an open secret:** Vidhu Gupta, Shauna M. Downs, Suparna Ghosh-Jerath, Karen Lock, and Archna Singh, "Unhealthy Fat in Street and Snack Foods in Low-Socioeconomic Settings in India: A Case Study of the Food Environments of Rural Villages and an Urban Slum," *Journal of Nutrition Education and Behavior*, April 2016.

167 **"The cheaper oil":** I interviewed Agarwal in his New Delhi office on October 27, 2017.

167 **Kamal Kapoor:** I interviewed Kapoor in his New Delhi office on November 1, 2017.

168 **1.38 billion people:** As of September 2020, http://worldpopulationreview .com/countries/india-population.

169 **In the five years:** EuroMonitor. https://www.euromonitor.com/usa.

169 **Domino's Pizza now operates:** https://biz.dominos.com/web/public /about-dominos/fun-facts.

169 **Subway has 508:** http://www.subway.com/en-in/exploreourworld.

169 **430 Pizza Huts:** http://www.yum.com/company/our-brands/pizza-hut.

169 **395 KFCs:** http://www.yum.com/company/our-brands/kfc.

169 **more than 300:** https://www.businesstoday.in/current/corporate/how -mcdonalds-is-combating-slowdown-through-its-expansive-menu-in-india /story/396058.html.

169 **have become a fixture:** Aakriti Gupta, Umesh Kapil, and Gajendra Singh, "Consumption of Junk Foods by School-aged Children in Rural Himachal Pradesh, India," *Indian Journal of Public Health* 62, no. 1 (2018). See also, Cheryl Tay, "'Junk food' Consumption in India a Growing Concern in Rural Areas, Research Reveals," *Food Navigator Asia*, April 18, 2018.

169 **Between 2014 and 2019:** EuroMonitor. https://www.euromonitor.com /usa.

169 **Sanjay Kumar:** I interviewed Kumar in his shop on October 25, 2017.

170 **growing their businesses:** David Stuckler, Martin McKee, Shah Ebrahim, and Sanjay Basu, "Manufacturing Epidemics: The Role of Global Producers in Increased Consumption of Unhealthy Commodities Including Processed Foods, Alcohol, and Tobacco," *PLoS Medicine* 9, no. 6 (June 2012).

170 **launched a health-focused:** Ratna Bhushan, "PepsiCo Takes 'Snack Smart' Logo Off Lays, Moves Away from Rice Bran Oil to Cut Costs," *Economic Times*, March 26, 2012.

170 **The move followed:** Mehmood Khan, MD, and George A. Mensah, MD, "Changing Practices to Improve Dietary Outcomes and Reduce Cardiovascular Risk: A Food Company's Perspective," IOM Heart Health. The memo has since been made public.

171 **daily intake recommended:** https://www.who.int/news-room/fact -sheets/detail/healthy-diet.

172 **485,756 metric tons:** According to its 2019 filing with the RSPO.

172 **plans to double:** Pearly Neo, "'In India, for India': PepsiCo Looks to Double Local Snacks Business," *Food Navigator Asia*, September 4, 2019.

172 **180,000 metric tons:** According to its 2019 filing with the RSPO.

172 **a spokesperson:** Email from Michael Maldonado, Global Media & External Communications, Yum! Brands, on December 1, 2017.

172 **92,534 metric tons:** According to its 2019 filing with the RSPO.

172 **a spokesperson:** Email from Andrea Abate, on August 24, 2018.

173 **After the signing:** Hawkes.

173 **one of the ten biggest:** "Against the Grain: Free Trade and Mexico's Junk Food Epidemic," GRAIN, February 2015.

173 **more than five-fold:** Figures according to IndexMundi. https://www .indexmundi.com.

173 **Whereas in 1980:** Andrew Jacobs and Matt Richtel, "A Nasty, NAFTA-Related Surprise: Mexico's Soaring Obesity," *New York Times*, December 11, 2017.

173 **eighty thousand lives:** Jacobs and Richtel.

173 **Today Mexico:** GRAIN, 2.

173 **a 2018 study:** Barry Popkin, "Obesity and the Food System Transformation in Latin America," *Obesity Review*, September 1, 2018, 9.

173 **"This is an equity issue":** I spoke with Heijnen by phone on August 2, 2017.

174 **some 20 percent:** "Instant Noodles: A Potential Vehicle for Micronutrient Fortification," Fortification Basics," USAID. https://www.dsm.com /content/dam/dsm/nip/en_US/documents/noodles.pdf.

174 **I visited:** I spent time in Jambi province on three separate occasions between May 2015 and July 2018.

174 **A 2019 study:** Ratna C. Purwestri, Bronwen Powell, Dominic Rowland, Nia N. Wirawan, Edi Waliyo, Maxi Lamanepa, Yusuf Habibie, and Amy Ickowitz, "From Growing Food to Growing Cash: Understanding the Drivers of Food Choice in the Context of Rapid Agrarian Change in Indonesia," *CIFOR InfoBrief*, July 2019.

174 **The rate of wasting:** Purwestri et al.

175 **Hassan Basri:** I interviewed Basri on July 21, 2018.

175 **Abdullah Sani:** I interviewed Sani on May 26, 2015.

176 **Remigio Caal:** I interviewed Caal in his village in the Petén on August 15, 2017.

177 **known for advancing:** Andrew Jacobs, "A Shadowy Industry Group

Shapes Food Policy Around the World," *New York Times*, September 16, 2019. (The International Life Sciences Institute responded to the *Times* article on its website: https://ilsi.org/ilsi-response-to-the-new-york-times.)

178 **a 20 percent tax:** Sanjay Basu, Kimberly Singer Babiarz, Shah Ebrahim, Sukurmar Vellakkal, David Stuckler, and Jeremy Goldhaber-Fiebert, "Palm Oil Taxes and Cardiovascular Disease Mortality in India: Economic-Epidemiologic Model," *British Medical Journal* 2013: 347.

178 **In 2015:** "Healthier Hawker Centre Business Model Catches on as Stallowners Switch to Healthier Oil and Salt," press release from Health Promotion Board, September 30, 2012.

178 **"There are some":** I interviewed Shauna Downs in her New Brunswick office on November 15, 2017.

178 **at a forum:** EAT Asia Pacific Food Forum 2017.

179 **press secretary:** Sathasivam's press secretary is Sivakumar Krishnan. I first emailed him to arrange a meeting on the morning of October 30, 2017. We met in the lobby later on that afternoon, after which he canceled my interview with the minister.

Chapter 8: Smog Over Singapore

180 **I'd been summoned:** I reported with Eyes on the Forest in July of 2018.

180 **an article:** Hans Nicholas Jong, "Indonesia to Investigate Death of Journalist Being Held for Defaming Palm Oil Company," *Mongabay*, June 21, 2018.

182 **the ability to award:** The Gecko Project and Mongabay, "Follow the Permits: How to Identify Corrupt Red Flags in Indonesian Land Deals," *Mongabay*, December 4, 2019.

182 **"We make a major step":** John M. Broder, "Bush Signs Broad Energy Bill," *New York Times*, December 19, 2007.

183 **Europe followed:** Susanne Retka Schill, "EU Adopts 10 Percent Biofuels Mandate," *Biodiesel Magazine*, January 1, 2009.

183 **Having long lobbied:** Rosidah Radzian, "The Impact of Renewable Fuel Standard (RFS2) on Palm Biodiesel's Market Access to the United States of America," Malaysian Palm Oil Board, 2011. http://palmoilis.mpob.gov.my /publications/OPIEJ/opiejv12n1-Rosidah.pdf. See also, Abrahm Lustgarten, "Palm Oil Was Supposed to Help Save the Planet. Instead It Unleashed a Catastrophe," *New York Times*, November 20, 2018.

183 **soared by 15 percent:** Eoin Bannon, "Cars and Trucks Burn Almost Half of All Palm Oil Used in Europe," *Transport & Environment*, May 31, 2016. See also, IndexMundi.com. 2011: 5,707,000 metric tons. 2012: 6,812,000 metric tons.

183 **second-largest importer:** Arthur Nelson, "Leaked Figures Show Spike in Palm Oil Use for Biodiesel in Europe," *The Guardian*, June 1, 2016.

183 **31 percent:** "Palm Oil Biofuels Market May See Shake-up in 2020, Heightening Leakage Risks," Chain Reaction Research, November 21, 2019.

183 **investigating a company:** Jong, "Indonesia to Investigate Death of Journalist."

184 **"All we got":** The Gecko Project, "The Making of a Palm Oil Fiefdom: The Story of Money, Power and Politics Behind the Devastation of a Forest-Rich District in Indonesian Borneo," October 11, 2019.

185 **the smoking capital:** Nathalia Tjandra, "Indonesia's Lax Smoking Laws Are Helping Next Generation to Get Hooked," *Jakarta Post*, June 4, 2018.

185 **surpassing that of Brazil:** Rhett A. Butler, "Despite Moratorium, Indonesia Now Has World's Highest Deforestation Rate," *Mongabay*, June 29, 2014.

185 **Some three-quarters:** K.G. Austin, A. Mosnier, J. Pirker, I. McCallum, S. Fritz, and P.S. Kasibhatla, "Shifting Patterns of Oil Palm Driven Deforestation in Indonesia and Implications for Zero-Deforestation Commitments," *Land Use Policy*, Vol. 69, December 2017.

186 **overtook Malaysia:** According to USDA statistics. https://ipad.fas.usda .gov/highlights/2007/12/indonesia_palmoil.

186 **"You once had":** I interviewed Hurowitz by phone on May 15, 2019, and on September 29, 2019.

186 **the corruption:** The Gecko Project, "How Corrupt Elections Fuel the Sell-off of Indonesia's Natural Resources," *Mongabay*, June 7, 2018.

186 **attempted to run:** The Gecko Project, "Abdon Nababan: 'North Sumatran Land Mafia Offered Me $21m to Win Election—and Then Hand Over Control of Government,'" *Mongabay*, June 21, 2018.

187 **a $250,000 bribe:** The Gecko Project, "Comment: It's Time to Confront the Collusion Between the Palm Oil Industry and Politicians That Is Driving Indonesia's Deforestation Crisis," *Mongabay*, April 18, 2018.

187 **the KPK brought:** Donal Fariz, "Battling Corruption in Indonesia's Elections," *The Diplomat*, May 15, 2019.

187 **gone unpunished:** Joshua Oppenheimer, "Why Today's Global Warming Has Roots in Indonesia's Genocidal Past," *The Guardian*, May 3, 2016.

188 **"A lot of what":** I traveled in Sumatra with this person (whose name I am leaving out for his safety) in November of 2016.

189 **as deep as sixty feet:** IPCC Special Report on Climate Change, Chapter 4, Land Degradation, 2018.

189 **And yet destroying:** Juka Miettinen, Chenghua Shi, and Soo Chin Liew, "Land Cover Distribution in the Peatlands of Peninsular Malaysia, Sumatra

and Borneo in 2015 with Changes Since 1990," *Global Ecology and Conservation* 6 (April 2016).

190 **One scam resulted:** Loren Bell, "Indonesia's Anti-corruption Agency Questions Former Minster of Forestry," *Mongabay*, November 21, 2014.

190 **17,400 square miles:** Hans Nicholas Jong, "Indonesia Forest-Clearing Ban Is Made Permanent, but Labeled 'Propaganda,'" *Mongabay*, August 14, 2019.

190 **more than 6 million acres:** Shannon N. Koplitz, Loretta J. Mickley, Miriam E. Marlier, Jonathan J. Buonocore, Patrick S. Kim, Tianjia Liu, Melissa P. Sulprizio, Ruth S. DeFries, Daniel J. Jacob, and Joel Schwartz, "Public Health Impacts of the Severe Haze in Equatorial Asia in September–October 2015: Demonstration of a New Framework for Informing Fire Management Strategies to Reduce Downwind Smoke Exposure," *Environmental Research Letters*, September 19, 2016.

190 **A study published:** P. Crippa, S. Castruccio, S. Archer-Nicholls, G.B. Lebron, M. Kuwata, A. Thota, S. Sumin, E. Butt, C. Wiedinmyer, and D.C. Spracklen, "Population Exposure to Hazardous Air Quality Due to the 2015 Fires in Equatorial Asia," *Scientific Reports*, November 16, 2016.

190 **"Hey, sorry":** I worked with the translator while reporting in Jambi in May of 2015 and have corresponded with him since.

192 **individual farmers:** G.C. Schoneveld, D. Ekowati, A. Andrianto, and S. van der Haar, "Modeling Peat- and Forestland Conversion by Oil Palm Smallholders in Indonesian Borneo," *Environmental Research Letters*, January 9, 2019.

192 **the KPK has found:** Mongabay staff, "Indonesia's Anti-graft Agency 'Eager to Intervene' in Palm Oil Sector," *Mongabay*, October 25, 2018.

196 **it's become clear:** Brad Plumer, "The EPA's Most Important Decision This Year Could Be over . . . Vegetable Oil?" *Washington Post*, April 27, 2012.

197 **massive amounts:** Thomas Guillaume, Martyna M. Kotowska, Dietrich Hertel, Alexander Knohl, Valentyna Krashevska, Kukuh Murtilaksono, Stefan Scheu, and Yakov Kuzyakov, "Carbon Costs and Benefits of Indonesian Rainforest Conversion to Plantations," *Nature Communications*, 2018.

197 **55,000 tons:** Miettinen et al.

197 **3,900 square miles:** Hans Nicholas Jong, "Indonesia Fires Emitted Double the Carbon of Amazon Fires, Research Shows," *Mongabay*, November 25, 2019. See also, Hans Nicholas Jong, "Indonesia Fires Cost Nation $5 Billion This Year: World Bank," *Mongabay*, December 20, 2019.

197 **a national park:** Dian Afriyanti, Lars Hein, Carolien Kroeze, Moham-

mad Zuhdi, and Asmadi Saad, "Scenarios for Withdrawal of Oil Palm Planta-tions from Peatlands in Jambi Province," *Regional Environmental Change*, Feb-ruary 28, 2019.

199 **a three-year moratorium:** Mongabay staff, "Indonesian President Signs 3-Year Freeze on New Oil Palm Licenses," *Mongabay*, September 20, 2018.

199 **aggressive biofuels mandates:** Hans Nicholas Jong, "'We've Been Negligent,' Indonesia's President Says as Fire Crisis Deepens," *Mongabay*, Sep-tember 17, 2019.

199 **In addition:** "Palm Oil Biofuels Market May See Shake-Up in 2020, Heightening Leakage Risks," Chain Reaction Research, November 21, 2019.

200 **as many as 13 million:** Chris Malins, "Biofuel to the Fire: The Impact of Continued Expansion of Palm and Soy Oil Demand Through Biofuel Poli-cy," Rainforest Foundation Norway, 2020. See also, Nithin Coca, "As Palm Oil for Biofuel Rises in Southeast Asia, Tropical Ecosystems Shrink," *Chinadia-logue*, April 20, 2020.

200 **could rival those:** Malins.

200 **the latter home:** David Smith, "Peat Bog as Big as England Found in Congo," *The Guardian*, May 27, 2014.

Chapter 9: Nutella and Other Smears

203 **"What kills the skunk":** Abraham Lincoln, quoted in Ida Minerva Tar-bell, *In the Footsteps of the Lincolns* (New York: Harper & Brothers, 1924), 380. Retrieved through Google Books.

203 **never actually aired:** Sarah Butler and Mark Sweney, "Iceland's Christ-mas TV Advert Rejected for Being Political," *The Guardian*, November 9, 2018.

203 **30 million times:** Magda Ibrahim, "Iceland's 'Rang-Tan' Ad Hits 30m Views Online and Prompts Petition," *PR Week*, November 13, 2018.

205 **158 percent:** According to IndexMundi. https://www.indexmundi .com/agriculture/?country=us&commodity=palm-oil&graph=domestic -consumption. 1981 = 104,000 metric tons. 1985 = 268,000 metric tons.

205 **responded with a media campaign:** Susan M. Martin, *The UP Saga* (Copenhagen: Nordic Institute of Asian Studies Press, 2003), 276.

205 **It featured:** Carole Sugarman, "A Slick War of Words," *Washington Post*, April 29, 1987.

205 **prompted protests:** Barbara Crossette, "International Report: Malaysia Opposes Labels on Palm Oil," *New York Times*, October 19, 1987.

205 **introduce a bill:** Sugarman.

206 **its own campaign:** Shakila Yacob, "Government, Business and

Lobbyists: The Politics of Palm Oil in US–Malaysia Relations," *The International History Review*, May 1, 2018: 921.

206 **also financed:** Steven Pratt, "World Grease War Slides into a Slugfest," *Chicago Tribune*, October 15, 1987.

206 **successfully petitioned:** Martin, *The UP Saga*, 277.

206 **But other adversaries:** Yacob.

208 **By 1989:** Martin, *The UP Saga*, 277.

208 **"American consumers":** Douglas C. McGill, "Tropical-Oil Exporters Seek Reprieve in U.S.," *New York Times*, February 3, 1989.

208 **The replacement:** "Tropical Fats Labeling: Malaysians Counterattack ASA Drive," *Journal of the American Oil Chemists Society* 64, no. 12 (1987): 1956, quoted in Nina Teicholz, *The Big Fat Surprise* (New York: Simon & Schuster, 2014), 236.

209 **efforts to defend:** Yacob, 910.

209 **an open letter:** CIFOR, "An Open Letter About Scientific Credibility and the Conservation of Tropical Forests," *Forest News*, October 26, 2010. See also, Rhett Butler, "Scientists Blast Greenwashing by Front Groups," *Mongabay*, October 27, 2010.

210 **World Growth International:** Alex Helan, "Greenwash and Spin: Palm Oil Lobby Targets Its Critics," *The Ecologist*, July 8, 2011. See also, Hanim Adnan, "Up Close with Pro-Palm Oil Lobbyist Alan Oxley," *The Star*, August 14, 2010.

210 **largest oil-palm plantation company:** According to the company's website. http://www.simedarbyplantation.com/corporate/overview/about-us.

210 **presented a proposal:** Ian Burrell and Martin Hickman, "Special Investigation: TV Company Takes Millions from Malaysian Government to Make Documentaries for BBC . . . About Malaysia," *The Independent*, August 17, 2011.

211 **"put a particular focus":** "Taib Paid Out $5 Million to Attack Sarawak Report!—International Expose," *Sarawak Report*, August 1, 2011.

211 **Jeffrey Sachs:** Ian Burrell, "Firm in BBC News-Fixing Row Targeted Poverty Guru," *The Independent*, November 17, 2011. (Sachs subsequently told Damian Carrington, of *The Guardian*, that he "never sought nor received a single penny from Sime Darby" and "would never serve as an 'ambassador' or 'champion' for Sime Darby." He said he did not know that the company was paying the FBC. See "Jeffrey Sachs Stung by the Corrosive Mix of Palm Oil and Publicity," *The Guardian*, November 17, 2011.)

211 **soon shut its doors:** Ian Burrell, "Company in News-Fixing Row Goes into Administration," *The Independent*, October 28, 2011.

211 **160,000 Italians:** Sarah Hucal, "The Italians Fighting Against 'an Invasion' of Palm Oil," *The Guardian*, December 9, 2015.

212 **For thirty days:** Hucal.

212 **Ferrero responded:** Niamh Michail, "Ferrero Defends Palm Oil in Nutella with Advert Against 'Unfair Smear Campaign,'" *Food Navigator*, December 5, 2017.

212 **released a report:** EFSA Panel on Contaminants in the Food Chain, "Risks for Human Health Related to the Presence of 3- and 2-monochloropropanediol (MCPD), and Their Fatty Acid Esters, and Glycidyl Fatty Acid Esters in Food," *EFSA Journal*, May 10, 2016.

212 **prompting companies:** Francesca Landini and Giancarlo Navach, "Nutella Maker Fights Back on Palm Oil After Cancer Risk Study," *Reuters*, January 11, 2017.

212 **dropped plans:** Sybille de La Hamaide, "French Parliament Scraps Planned Extra Tax on Palm Oil," *Reuters*, June 23, 2016.

213 **made it clear:** Arthur Neslen and Joe Sandler Clarke, "French Politicians Scrapped Palm Oil Tax After Indonesia Execution Warning," *DeSmog*, March 18, 2019.

213 **"We are legislating":** Niamh Michail, "French MPs Drop Palm Oil Tax—but Accuse Producer Countries of Blackmail," *Food Navigator*, June 26, 2016.

213 **submitted a letter:** Loren Bell, "139 Scientists Shoot Down 'Misleading' Reports from Malaysia Peat Congress," *Mongabay*, October 4, 2016.

213 **more of the same:** A. Ananthalaksmi and Emily Chow, "Fearing Tobacco's Fate, Palm Oil Industry Fights Back," *Reuters*, August 21, 2019.

214 **determined that its creator:** Jake Tapper and Max Culhane, "Al Gore YouTube Spoof Not So Amateurish," *ABC News*, August 5, 2006.

214 **The firm specializes:** "DCI Group Background," *DeSmog*. https://www.desmogblog.com/dci-group.

214 **among its clients:** Dave Levinthal, "Lobbying Firm Fires 12," *Politico*, April 17, 2012.

214 **Burmese military junta:** Marc Ambinder, "The DCI Group Responds on Burma," *The Atlantic*, May 19, 2008.

214 **published an online magazine:** Mike McIntire, "Odd Alliance: Business Lobby and Tea Party," *New York Times*, March 30, 2011.

214 **news outlets reported:** Ananthalaksmi and Chow, of Reuters.

215 **took out several ads:** Joe Sandler Clarke, "How Palm Oil Sparked a

Diplomatic Row Between Europe and Southeast Asia," *Unearthed*, March 18, 2009.

215 **"Attempting to reason":** Ananthalaksmi and Chow.

215 **have long complained:** Hans Nicholas Jong, "Indonesian Oil Palm Smallholders Sue State over Subsidy to Biofuel Producers," *Mongabay*, April 24, 2018.

216 **$6 a day:** Laura Villadiego, "Precarious Employment, Lower Pay and Exposure to Chemicals: The Gender Divide in the Palm Oil Industry," *Equal Times*, July 3, 2017.

216 **closer to $9:** Malay Mail staff, "Migrant Workers in Oil Palm Plantations Deserve Better Treatment—Tenaganita," *Malay Mail*, September 25, 2018. 38.46 Malaysian ringgit = $9.22.

216 **rank among the wealthiest:** Figures according to "Malaysia's Richest" and "Indonesia's Richest," *Forbes*, October 2020. See also, *"Tycoons in the Indonesian Palm Oil Sector 2018,"* TuK Indonesia. https://www.tuk.or.id/2019/03/08/tycoon-in-the-indonesian-palm-oil/?lang=en.

216 **Fife told me:** I spoke with Fife by phone on November 29, 2017. Dr. Bruce Fife, *The Palm Oil Miracle* (Colorado Springs: Picadilly Books, 2007).

216 **closed abruptly:** Stan Diel, "Birmingham-Based Internet College to Close, Blames Economy," *Alabama.com*, July 10, 2010.

217 **class-action lawsuit:** Stan Diel, "Former Clayton College Students to Get up to $2.31 Million, Tuition Discounts," *Alabama.com*, November 18, 2011.

217 **an article:** Sowmya Kadandale et al., "Palm Oil Industry and Noncommunicable Diseases," *World Health Organization Bulletin*, January 2019.

218 **released a statement:** Bernama, "Council Accuses WHO of 'Cherry-Picking' in Study Against Palm Oil," *FMT News*, January 10, 2019.

218 **"The eco-colonialists":** Ananthalaksmi and Chow.

218 **a report on diet:** "Diet, Nutrition and the Prevention of Chronic Diseases: Report of the Joint WHO/FAO Expert Consultation," WHO Technical Report Series, No. 916.

218 **Yach said:** I have spoken with Yach several times since 2016. In May of 2019, I met him at the Midtown Manhattan offices of Foundation for a Smoke-Free World, where he now serves as president.

219 **sent a letter:** Emily Chow and Joseph Sipalan, "Malaysia Could Curb French Purchases If Palm Oil Use Restricted," *Reuters*, January 23, 2019.

219 **decreed the biofuels decision:** "Teresa Kok Blasts EU's Palm Oil Decision Based on 'Politics of Protectionism,'" *The Star*, May 15, 2019.

219 **"toxic entities":** A. Ananthalakshmi and Mei Mei Chu, "Malaysian

Palm Oil Bosses Urge Action Against 'Toxic' Environmental Groups," *Reuters*, February 4, 2020.

220 **had been deported:** Richard C. Paddock, "American Journalist Is Arrested in Indonesia Over Visa Issue," *New York Times*, January 22, 2020.

Chapter 10: Fight the Power

221 **"We can't save":** Greta Thunberg, TED Talk, May 2018.

221 **The team's leader:** I spoke with Henry by phone on July 31, 2019. I spoke with Maya Marewu, the Indonesian activist who boarded the *Stolt*, on July 30, 2019.

222 **published a report:** "The Final Countdown: Now or Never to Reform the Palm Oil Industry," Greenpeace International, September 19, 2018. Available on the organization's website.

224 **"As awful a place":** Kevin Barry, *Night Boat to Tangier* (New York: Doubleday, 2019), 1.

224 **My introduction:** I spent time with the RAN activists from March 29 to March 31, 2015.

225 **launched an initiative:** "The Snack Food 20," Rainforest Action Network. https://www.ran.org/sf20scorecard.

226 **Lindsey Allen:** Allen has since left the organization.

226 **The two boycotted:** Hillary Rosner, "Palm Oil and Scout Cookies—the Battle Drags On," *New York Times*, February 13, 2002.

227 **"I think it's time":** Elizabeth Shogren, "Two Scouts Want Palm Oil Out of Famous Cookies," *NPR*, July 4, 2011.

227 **twelve of its sixteen members:** According to the RSPO website. https://www.rspo.org.

228 **The group's annual meetings:** I have attended RSPO meetings in Kuala Lumpur, Bangkok, Milan, and London.

228 **released a report:** "Who Watches the Watchmen?" published by the EIA on November 16, 2016. https://eia-international.org/report/who-watches-the-watchmen.

229 **the resignation:** Mongabay staff, "PanEco Resigns from RSPO Over 'Sheer Level of Inaction,'" *Mongabay*, June 3, 2016.

230 **withdrew its membership:** Reuters staff, "Liberia's Biggest Palm Oil Project Quits Eco-Certification Scheme," *Reuters*, July 21, 2018.

230 **a second report:** "Who Watches the Watchmen 2," EIA.

230 **19 percent:** According to the RSPO website.

230 **issued a report:** Rhett A. Butler, "Palm Oil Giant Profiting Off Tiger Habitat Destruction, Alleges Greenpeace," *Mongabay*, October 22, 2013.

230 **twenty-three "dirty" palm oil companies:** Mongabay staff, "Norway's Wealth Fund Dumps 23 Palm Oil Companies Under New Deforestation Policy," *Mongabay*, March 11, 2013.

231 **might be a chance:** Based on conversations with Hurowitz. See also, Nathanael Johnson, "48 Hours That Changed the Future of Rainforests," *Grist.com*, April 2, 2015.

233 **an open letter:** Jeremy Hance, "Scientists Say Massive Palm Oil Plantation Will 'Cut the Heart Out' of Cameroon's Rainforest," *Mongabay*, March 15, 2012.

233 **finally suspended:** Christiane Badgley, "When Wall Street Went to Africa," *Foreign Policy*, July 11, 2014.

235 **pulled out:** Front Page Africa staff, "Sime Darby Finally Shuts Down; Leaves Liberia," *Front Page Africa*, January 17, 2020.

235 **Global Witness found:** "Defending Tomorrow," released by Global Witness in July 2020. https://www.globalwitness.org/en/campaigns/environmental -activists/defending-tomorrow.

235 **shot dead:** Human Rights Watch, "Thailand Land Rights Activist Gunned Down," *On Dangerous Ground*, 8. https://www.hrw.org/news/2015/02 /14/thailand-land-rights-activist-gunned-down.

235 **was killed:** Action Aid, "Guatemalan Activist Murdered Protesting Chemical Leak at Palm Oil Plant," September 2015. https://www.actionaidusa .org/2015/09/guatemalan-activist-murdered-protesting-chemical-leak-palm -oil-plant.

236 **Bill Kayong was shot:** Borneo Post staff, "PKR Miri Branch Secretary Bill Kayong Shot Dead in Drive-by," *The Borneo Post*, June 21, 2016.

236 **In 2012:** Jocelyn C. Zuckerman, "The Violent Costs of the Global Palm Oil Boom," *newyorker.com*, December 10, 2016.

236 **two Medan-based journalists:** Ayat S. Karokaro, "Indonesian Journalists Critical of Illegal Palm Plantation Found Dead," *Mongabay*, November 4, 2019. See also, Richard C. Paddock, "A Hard-Fighting Indonesian Lawyer's Death Has Colleagues Asking Questions," *New York Times*, October 24, 2019.

237 **larger than Singapore:** "Indonesian Forest Fires Crisis: Palm Oil and Pulp Companies with Largest Burned Land Areas Are Going Unpunished," Greenpeace Southeast Asia, September 24, 2019.

237 **credibly accused:** "Large Scale Bribery and Illegal Land-Use Violations

Alleged on Large Parts of Golden Agri-Resources Palm Oil Plantations," Forest Peoples Programme, March 20, 2020.

238 **IndoAgri withdrew:** "Indofood Withdraws from RSPO," *Oils & Fats International*, February 11, 2019.

239 **made a public statement:** Fink's letter, "Profit & Purpose," is available on the company's website.

240 **the trade press:** MI News Network, "Photos: Six Greenpeace Activists Arrested on Board Ship Loaded with Palm Oil," *Shipping News*, November 20, 2018.

Epilogue: Post-Pandemic Palm

243 **Between 60 and 75 percent:** "Neglected Tropical Diseases," World Health Organization. https://www.who.int/neglected_diseases/diseases/zoonoses/en.

243 **the number:** Kate E. Jones, Mikkita G. Patel, Marc A. Levy, Adam Storeygard, Deborah Balk, John L. Gittleman, and Peter Daszak, "Global Trends in Emerging Infectious Diseases," *Nature* 451, 990–993, 2008.

243 **A third of these:** Simon L. Lewis, David P. Edwards, and David Galbraith, "Increasing Human Dominance of Tropical Forests," *Science* 349, no. 6250 (August 22, 2015): 827–32.

244 **"Our consumption drives":** I spoke to Daszak by phone on March 22, 2020.

244 **half of the world's:** The Oxford Martin Programme on Global Development, University of Oxford, "Agricultural Land by Global Diets," 51 million square kilometers. World Bank: 48.6 square kilometers in 2016.

244 **some 850 million:** *The State of Food Security and Nutrition in the World 2019*, Food and Agriculture Organization of the United Nations.

244 **One in three:** According to the World Health Organization. https://www.who.int/news-room/fact-sheets/detail/obesity-and-overweight.

244 **has reported:** IPCC, "Special Report: Global Warming of 1.5°: Summary for Policymakers."

244 **some 8 percent:** David Gibbs, Nancy Harris, and Frances Seymour, "By the Numbers: The Value of Tropical Forests in the Climate Change Equation," World Resources Institute, October 4, 2018.

244 **a mass extinction:** Rachel Nuwer, "Mass Extinctions Are Accelerating, Scientists Report," *New York Times*, June 1, 2020.

245 **We can start:** I was guided in this section by the following reports,

among others: "Covid 19: Urgent Call to Protect People and Nature," published by WWF in Spring of 2020; Timothy D. Searchinger, Chris Malins, Patrice Dumas, David Baldock, Joe Glauber, Thomas Jayne, Jikum Huang, and Paswel Marnya, "Revising Public Agricultural Support to Mitigate Climate Change," published by the World Bank Group in 2020; "The 2020 Global Nutrition Report," Development Initiatives Poverty Research, Ltd, Bristol, U.K.; Timothy Searchinger, Richard Waite, Craig Hanson, and Janet Ranganathan, "Creating a Sustainable Food Future: A Menu of Solutions to Feed Nearly 10 Billion People by 2050," World Resources Institute, 2019; and The IPCC Special Report on Climate Change, issued in August of 2019.

245 **the leading cause:** "The 2020 Global Nutrition Report." Development Initiatives Poverty Research, Ltd, Bristol, U.K., 16. https://globalnutritionreport.org/reports/2020-global-nutrition-report.

246 **more vulnerable to COVID-19:** According to the Centers for Disease Control and Prevention. https://www.cdc.gov/coronavirus/2019-ncov/need-extra-precautions/people-with-medical-conditions.html.

246 **30 billion tons:** CarbonBrief, "The Carbon Brief Profile, Indonesia." https://www.carbonbrief.org/the-carbon-brief-profile-indonesia.

246 **7.5 million acres:** Hannah V. Cooper, Stephanie Evers, Paul Aplin, Neil Crout, Mohd Puat Bin Dahalan, and Sofie Sjogersten, "Greenhouse Gas Emissions Resulting from Conversion of Peat Swamp Forest to Oil Palm Plantation," *Nature Communications* 11, 407 (January 2020). See also, Sophie Sjogersten, "Palm Oil: Research Shows That New Plantations Produce Double the Emissions of Mature Ones," *The Conversation*, January 22, 2020. I converted 3 million hectares to 7.4 million acres.

246 **breeding programs:** For more on this, see Anuradha Raghu, "New Dwarf Trees Set to Revolutionize Palm Oil Market," *BNN Bloomberg*, October 3, 2018.

247 **protect 80 percent:** "Indigenous Peoples: The Unsung Heroes of Conservation," United Nations website. See also The World Bank, "Indigenous Peoples."

247 **a program initiated:** David M. Lapola, Luiz A. Martinelli, Carlos A. Peres, Jean P.H.B. Ometto, et al., "Persuasive Transition of the Brazilian Land-Use System," *Nature Climate Change* 4 (December 20, 2013). See also, Wayne S. Walker, Seth R. Gorelik, Alessandro Baccini, Jose Luis Aragon-Osejo, Carmen Josse, Chris Meyer, Marcia N. Macedo, Cicero Augusto, Sandra Rios, Tuntiak Katan, Alana Almeida de Souza, Saul Cuellar, Andres Llanos, Irene Zager, Gregorio Díaz Mirabal, Kylen K. Solvik, Mary K. Farina, Paulo Moutinho, and Stephan Schwartzman, "The Role of Forest Conversion, Degradation, and Disturbance in the Carbon Dynamics of Amazon Indigenous Territories and

Protected Areas," *Proceedings of the National Academy of Sciences*, February 11, 2020.

247 **creating a library:** John Cumbers, "There Is More Money in the Borneo Rainforest's Biodiversity Than in Its Deforestation," *Forbes*, September 12, 2019.

248 **could follow the lead:** Carol J. Clouse, "The U.N.'s Grand Plan to Save Forests Hasn't Worked but Some Still Believe It Can," *Mongabay*, July 14, 2020. See also, Hans Nicholas Jong, "Indonesia to Get First Payment from Norway Under $1Bn REDD+ Scheme," *Mongabay*, February 20, 2019, and "Norway to Complete $1 Billion Payment to Brazil for Protecting Amazon," *Reuters*, September 15, 2015.

248 **called on companies:** "230 Investors with $16.2 trillion in AUM Call for Corporate Action on Deforestation, Signaling Support for the Amazon," *Principles for Responsible Investment*, September 18, 2019.

248 **more than sixty companies:** Michael Taylor, "Norway's Wealth Fund Ditches 33 Palm Oil Firms Over Deforestation," *Reuters*, February 28, 2019.

249 **a recent pledge:** Larry Fink, "A Fundamental Reshaping of Finance," BlackRock, https://www.blackrock.com/corporate/investor-relations/larry-fink-ceo-letter.

249 **some $7 trillion:** Andrew Ross Sorkin, "BlackRock C.E.O. Larry Fink: Climate Crisis Will Reshape Finance," *New York Times*, January 14, 2020.

249 **orangutan video:** According to Simon Lord, former chief sustainability officer for Sime Darby.

250 **Tom Kelleher and Tom Jeffries:** I interviewed both men in Madison, Wisconsin, on February 10 and 11, 2020.

250 **capable of producing:** Bonnie A. McNeil and David T. Stuart, "*Lipomyces starkeyi*: An Emerging Cell Factory for Production of Lipids, Oleochemicals and Biotechnology Applications," *World Journal of Microbiology and Biotechnology* 34, 147 (2018).

252 **$20 million:** Akshat Rathi, "Bill Gates–Led Fund Invests in Synthetic Palm Oil Startup," *Bloomberg Green*, March 2, 2020. See also, Jeffrey J. Bussgang and Olivia Hull, "C16 Biosciences: Lab-Grown Palm Oil," Harvard Business School case study, November 12, 2019.

252 **researchers at the University of Bath:** "Developing an Alternative to Palm Oil from Waste Resource, Using Yeast," from the website of the Department of Chemical Engineering, University of Bath.

252 **Tom Kelleher's daughter:** I interviewed Nicole Kelleher in Madison, Wisconsin, on February 10 and 11.

252 **She had calculated:** Kelleher came to her figure as follows: According to the International Council on Clean Transportation, palm oil–related carbon contributions from tropical deforestation total roughly 700 million metric tons annually. (https://theicct.org/blog/staff/palm-oil-elephant-greenhouse). According to the *Journal of the Institute of Food Science and Technology*, the cosmetics and beauty sector accounts for two percent of the total palm oil market. (https://fstjournal.org/features/33-1/sustainable-palm-oil). Two percent of 700 million metric tons of carbon = 14 million metric tons. Kelleher assumes that Sarnaya will use palm oil–free glycerol, a waste product, as its carbon source.

253 **When it comes:** For more on this, see Sophie Parsons, Sofia Raikova, and Christopher J. Chuck, "The Viability and Desirability of Replacing Palm Oil," *Nature Sustainability* 3 (March 9, 2020): 412–18.

253 **In early June:** Anna Russell, "How Statues in Britain Began to Fall," *newyorker.com*, June 22, 2020.

254 **statue of King Leopold II:** Monika Pronczuk and Mihir Zaveri, "Statue of Leopold II, Belgian King Who Brutalized Congo, Is Removed in Antwerp," *New York Times*, June 9, 2020.

254 **By June 12:** Jamie Bowman, "Calls for Debate on Bolton Park Name and Lord Leverhulme's Slave Labour Links," *Bolton News*, June 12, 2020.

254 **This particular site:** I visited Lusanga, where I interviewed Martens, Ngongo, and several of the artists, from February 14 to February 19, 2019.

255 **back in 1924:** See Chapter 3 in this book.

256 **the *Times* named:** Roberta Smith, Holland Cotter, and Jason Farago, "The Best Art of 2017," *New York Times*, December 6, 2017. See also, Randy Kennedy, "Chocolate Sculpture, with a Bitter Taste of Colonialism," *New York Times*, February 2, 2017.

257 **palm oil communities:** Hans Nicholas Jong, "Seeking Justice Against Palm Oil Firms, Victims Call Out Banks Behind Them," *Mongabay*, October 10, 2019. See also, Chris Arsenault, "Congo Plantation Firm Financed by U.K. Aid Accused of Breaking Promise to Help Workers," *Reuters*, February 28, 2017; Claire Provost, "Farmers Sue World Bank Lending Arm Over Alleged Violence in Honduras," *The Guardian*, March 8, 2017; and Hans Nicholas Jong, "Indonesian Oil Palm Smallholders Sue State over Subsidy to Biofuel Producers," *Mongabay*, April 24, 2018.

PERMISSIONS

52 Jocelyn Zuckerman
56 Edson Varas. Used with permission of Ayrson Heráclito.
63 Topical Press Agency / Stringer
99 Rhett A. Butler / Mongabay
103 Jocelyn Zuckerman
104 Jocelyn Zuckerman
111 Paul Hilton
114 The World Factbook
120 Paul Hilton
125 Paul Hilton
128 Paul Hilton
130 Paul Hilton
131 Paul Hilton
138 Courtesy of STAS
148 Marco Di Lauro
152 Jason Motlagh
154 Luciana Téllez / Human Rights Watch
166 Jocelyn Zuckerman
171 Jocelyn Zuckerman
177 Jocelyn Zuckerman
182 Kemal Jufri
189 Rhett A. Butler / Mongabay
195 Jocelyn Zuckerman
205 Greenpeace
207 Courtesy of Karen Sokolof Javitch
212 Jeff Koehler
223 Greenpeace
234 Goldman Environmental Prize
238 Rainforest Action Network
251 Courtesy of Tom Kelleher
258 Léonard Pongo

INDEX

ABOUT THE AUTHOR

Jocelyn C. Zuckerman is the former deputy editor of *Gourmet*, articles editor of *OnEarth*, and executive editor of *Modern Farmer*. An alumna of Columbia University's Graduate School of Journalism and a former fellow with the Washington, DC–based Alicia Patterson Foundation, she has written for *Fast Company*, the *American Prospect*, *Vogue*, and many other publications. She lives in Brooklyn, with her husband and two children.

PUBLISHING IN THE PUBLIC INTEREST

Thank you for reading this book published by The New Press. The New Press is a nonprofit, public interest publisher. New Press books and authors play a crucial role in sparking conversations about the key political and social issues of our day.

We hope you enjoyed this book and that you will stay in touch with The New Press. Here are a few ways to stay up to date with our books, events, and the issues we cover:

- Sign up at www.thenewpress.com/subscribe to receive updates on New Press authors and issues and to be notified about local events
- Like us on Facebook: www.facebook.com/newpressbooks
- Follow us on Twitter: www.twitter.com/thenewpress
- Follow us on Instagram: www.instagram.com/thenewpress

Please consider buying New Press books for yourself; for friends and family; or to donate to schools, libraries, community centers, prison libraries, and other organizations involved with the issues our authors write about.

The New Press is a 501(c)(3) nonprofit organization. You can also support our work with a tax-deductible gift by visiting www.thenewpress.com/donate.